Fish and Invertebrate Culture

FISH AND INVERTEBRATE CULTURE

Water Management in Closed Systems

STEPHEN SPOTTE

Director of Aquariums
Aquarium Systems, Inc.
Eastlade, Ohio

Wiley–Interscience, a Division of John Wiley & Sons, Inc.

New York • London • Sydney • Toronto

Library of Congress Catalog Card Number: 74-130433

ISBN   0   471   81760   0

Printed in the United States of America

10   9   8   7   6

For Sara

Foreword

For more than four thousand years men have kept fish alive in ponds or tanks for food, and at least one species has been bred regularly in captivity since before the Christian Era. Fish culture and the keeping of fishes as pets have been practiced more or less continuously since the times of the ancient Romans and Chinese, and some of the methods used today are virtually the same as those devised centuries ago. Keepers of fish have been much slower to take advantage of the benefits of science and technology than have their counterparts, the farmer and animal husbandman. Except for an eighteenth-century flurry of interest in the wonders of fertilization and early embryology —which eventually led to the artificial incubation universally practiced in trout hatcheries—fish culturists, fanciers, and professional aquarists have been content to go their traditional, empirical ways. The beginnings of a systematic application of the scientific method to solve piscicultural problems are still within the memory of living men, and the methodical adoption of technological advances from other fields is hardly any older.

This has been especially true of the maintenance and control of the water in which captive fishes must live. So poor has been the fish man's understanding of the complicated interactions between the fish and its watery element that he has failed to recognize bad water as the underlying cause of most of his failures. Man's lack of understanding of the fish's way of life stems primarily from the fact that he is a terrestrial animal, isolated from his desiccating, hostile environment, while the fish, still bathed in the fluid in which life itself arose, maintains the most intimate relationship with its surroundings, a relationship whose sensitivity goes back to the very origin of life. Evolution has provided fishes with limited homeostatic mechanisms, which are all too often overtaxed by the conditions they find in goldfish bowls and hatchery troughs alike.

Nevertheless, by locating his hatcheries and fishponds near unlimited supplies of water, or by dealing only with species preadapted to polluted environments, the fish culturist or fancier has managed to be successful enough in the past, in spite of his ignorance. But water is fast becoming a scarce commodity, and in both Europe and America trout are now reared in recirculating water. Moreover, the exigencies of mariculture and the mass-culturing of new and more delicate kinds of fishes and invertebrates demand a degree of water control that old-time fish men would never have dreamed about.

Now, for the first time, people who want to raise aquatic animals as economically as possible, or need to keep them alive in captivity for any reason (in aquarium or pond, at home or in the laboratory), have a book that not only will tell them what to do but will explain the reasons for doing it. Now, fish culturists, experimental biologists, and amateur and professional aquarists have a firm bridge between their fish-keeping activities and the diverse chemical, biological, limnological, and oceanographic literature that bears on them. The time for a strictly scientific aquaculture has come; this pioneering book is going to lead the way.

JAMES W. ATZ

The American Museum of Natural History
New York City
May, 1970

Preface

In recent years there has been a worldwide upsurge of interest in maintaining freshwater and marine fishes and invertebrates in captivity. Aquarists have traditionally been interested in these animals from the standpoint of exhibition. But much of today's enthusiasm can be traced to the increased use of aquatic animals in research and also to the burgeoning science of aquaculture, in which marketable aquatic species are farmed on a commercial scale. In any case, successful results depend on sustaining and reproducing stable water conditions.

This book deals with the closed-system approach to culturing aquatic animals, in which the water is recycled and used again. In semiclosed and open systems, water is continuously discarded and replenished from a natural source. Mastery of closed systems has several advantages. It enables the culturist to raise marine organisms hundreds of miles from the sea, the advantage being that site selection for a saltwater laboratory, hatchery, or public aquarium need not be dictated by an available supply of natural sea water. Synthetic sea water can be used instead. Both freshwater and seawater closed systems can be operated in heavily polluted areas without depending on the local water supply. But most important, closed systems offer better environmental control. Natural water is subject to seasonal temperature fluctuations and may sometimes carry disease organisms, silt, pollutants, and undesirable animals that compete with the cultured specimens for space and nutrients or even prey upon them.

Aquatic animals change the chemistry of their culture water. Their physiology is in turn affected by these changes, usually in adverse ways. The purpose of this book is to show how to culture fishes and invertebrates in closed systems by controlling the chemical factors in the environment bringing about changes in normal physiology. The major concept presented is that accumulating toxic metabolites are the primary limiting factors in aquatic animal culture and that removing them on a continuous basis should be the foremost objective in routine water management.

Previous books dealing with culture problems of aquatic animals have emphasized a biological viewpoint. Osmotic imbalance, thermal shock, partial asphyxiation, lowered disease resistance, stunting, and loss of fecundity are typical problems of captivity, but they can seldom be corrected by treating the biological effects. These and many other maladies are often the tangible results of deteriorating water quality.

This book has been designed as a water management handbook for the researcher, teacher, or advanced student maintaining a modest collection of animals in the laboratory. It is also written for the fishery biologist and hatchery manager responsible for thousands of animals in various stages of development. The size of a culture system has little bearing on the isolation and solution of water management problems. These are essentially the same in a 10-gallon aquarium or a 10-million-gallon hatchery.

The term "aquatic animals," as used throughout the book, refers collectively to fishes and invertebrates of both freshwater and marine origins. In general, the same water quality standards and water management techniques apply to both groups of animals and to both types of environment.

The book is divided into two sections. Part I concentrates on the unfavorable effects that animals have on captive water. Part II deals with the effects of the culture water on the physiology of the animals. Both parts stress theoretical principles as well as practical water management techniques. It is assumed that the reader will have a stronger background and orientation in biology than in chemistry. Therefore, only theoretical data relating to the practical aspect of the subject are presented. Rigorous thermodynamic and mathematical functions of important reactions are avoided because they serve no practical purpose. Laboratory tests in the last chapter have been simplified for the same reason. Finally, the literature cited is meant to be representative of published work, rather than a complete survey.

STEPHEN H. SPOTTE

Niagara Falls, New York
May, 1970

Acknowledgments

I am grateful to the following people for reading all or parts of the original manuscript and for offering their comments.

JAMES W. ATZ
Curator and Dean Bibliographer
The American Museum of Natural History
New York, New York

ROGER E. BURROWS
Director, Salmon-Cultural Laboratory
U. S. Department of the Interior
Bureau of Sport Fisheries and Wildlife
Longview, Washington

VERNON C. GOLDIZEN
Program Director for Invertebrate Studies
Aquatic Sciences, Inc.
Boca Raton, Florida

WILLIAM J. HARGIS, JR.
Director, Virginia Institute of Marine Science
Gloucester Point, Virginia

GLENN L. HOFFMAN
Parasitologist
Eastern Fish Disease Laboratory
U. S. Department of the Interior
Bureau of Sport Fisheries and Wildlife
Kearneysville, West Virginia

Frank J. Kadell
Director of Engineering
Aquarium Systems, Inc.
Eastlake, Ohio

Justus F. Muller
Editor, Journal of Parasitology
Professor, Department of Microbiology
S.U.N.Y. Upstate Medical Center
Syracuse, New York

I am particularly grateful to Barbara E. Gailey, who prepared the original figures and adapted others from previously published works, to Bruce R. Powers, Assistant Professor of English, Niagara University, for editorial assistance, and to Roberta E. O'Donnell for typing the manuscript.

My thanks also go to the publishers, companies and authors who allowed me to reprint portions of their copyrighted work.

S. H. S.

Contents

PART I   EFFECTS OF ANIMALS ON CAPTIVE WATER

Chapter 1   Biological Filtration   **3**

1.1   Definition and Function,   3
1.2   Care of the Filter Bed,   5
1.3   Determination of Carrying Capacity,   15
1.4   The "Conditioned" System,   17
1.5   Mechanisms of Detritus Formation   20

Chapter 2   Mechanical Filtration   **22**

2.1   Definition and Function,   22
2.2   Mechanisms of Mechanical Filtration by Gravel,   22
2.3   The Biological Gravel Bed as a Mechanical Filter,   24
2.4   Rapid Sand Filters,   25
2.5   Diatomaceous Earth Filters,   28
2.6   Factors Affecting the Efficiency of DE Filters,   29
2.7   Troubleshooting Clogged Filter Sleeves,   35
2.8   Evaluation of Methods   38

Chapter 3   Chemical Filtration   **40**

3.1   Definition and Function,   40
3.2   Removal of Dissolved Organics by Adsorption,   41
3.3   Removal of Dissolved Organics by Foam Fractionation,   52
3.4   Removal of Dissolved Organics by Oxidation   56
3.5   Evaluation of Methods,   58

Chapter 4   The Carbon Dioxide System   **60**

4.1   Definition of Terms,   60
4.2   Derivation of Carbonate and Bicarbonate Ions,   65

4.3   Factors Affecting the Solubility of Mineral Carbonate,   66
4.4   Maintaining pH: Factors Causing a Gradual Decline in pH,   69
4.5   The Significance of pH   71

PART II   EFFECTS OF CAPTIVE WATER ON ANIMALS   73

Chapter 5   Respiration   75

5.1   Factors Affecting Oxygen Solubility,   75
5.2   Temperature,   78
5.3   Other Factors Affecting Respiration,   81

Chapter 6   Salts and Elements   84

6.1   Salinity, Chlorinity, and Specific Gravity,   84
6.2   Functions and Uptake of Elements,   85
6.3   Toxic Effects of Elements,   89
6.4   Synthetic Sea Water,   93

Chapter 7   Toxic Metabolites   102

7.1   Nature of the Effects,   102
7.2   Ammonia,   102
7.3   Pheromones,   107

Chapter 8   Disease Prevention by Environmental Control   109

8.1   Immunity and the Environment,   109
8.2   Disease Prevention and the Environment,   111
8.3   Treating Diseases,   113

Chapter 9   Laboratory Tests   115

9.1   Ammonia (as Total NH_4^+),   115
9.2   Nitrite and Nitrate,   116
9.3   Measurement of Salinity and Specific Gravity,   117
9.4   Dissolved Oxygen   117

Credits for Illustrative Material   121

Literature Cited   123

Index   133

PART I

Effects of Animals on Captive Water

CHAPTER 1

Biological Filtration

1.1 DEFINITION AND FUNCTION

Three types of filtration are used in closed system culturing: biological, mechanical, and chemical. Of these, biological filtration is the most important.

Biological filtration is defined as the mineralization of organic nitrogenous compounds, nitrification, and denitrification by bacteria suspended in the water and attached to the gravel in the filter bed.

Heterotrophic and autotrophic bacteria are the major groups present in culture systems. Heterotrophic species utilize organic nitrogenous compounds excreted by the animals as energy sources and convert them into simple compounds, such as ammonia. The mineralization of these organics is the first stage in biological filtration. It is accomplished in two steps: *ammonification*, which is the chemical breakdown of proteins and nucleic acids, producing amino acids and organic nitrogenous bases; and *deamination*, in which a portion of the organics and some of the products of ammonification are converted to inorganic compounds. An example of the latter is the breakdown of urea (eq. 1) to produce carbon dioxide and un-ionized ammonia. (This reaction can also proceed as a purely chemical process, but the deamination of amino acids and related substances requires the presence of bacteria.)

$$O=C\begin{array}{c} \diagup NH_2 \\ \diagdown NH_2 \end{array} + H_2O \rightarrow CO_2 + 2NH_3 \qquad (1)$$

Once organics have been mineralized by heterotrophs, biological filtration shifts to the second stage, which is nitrification. *Nitrification* is the biological oxidation of ammonia to nitrite and of nitrite to nitrate by autotrophic bacteria, as illustrated in Fig. 1. These organisms, unlike heterotrophs, require an inorganic substrate as an energy source and utilize carbon dioxide as their only source of carbon.

Nitrosomonas sp. and *Nitrobacter sp.* are the principal nitrifying bacteria in culture systems. *Nitrosomonas* oxidizes ammonia to nitrite (eq. 2). *Nitrobacter* oxidizes nitrite to nitrate (eq. 3). Both reactions show a fall in free

3

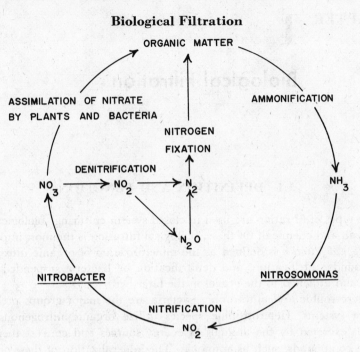

Figure 1. The nitrogen cycle.

energy. The significance of reactions 2 and 3 is a conversion of toxic ammonia to nitrate.

$$NH_4^+ + OH^- + 1.5O_2 \rightarrow H^+ + NO_2^- + 2H_2O \qquad (2)$$
$$\Delta G^\circ = -59.4 \text{ kcal}$$

$$NO_2^- + 0.5O_2 \rightarrow NO_3^- \qquad (3)$$
$$\Delta G^\circ = -18.0 \text{ kcal}$$

The third, and last, stage in biological filtration is *denitrification*. This process is defined by Vaccaro (1965) as a biological reduction of nitrate or nitrite to either nitrous oxide or free nitrogen (Fig. 1). Denitrification can apparently be carried out by both heterotrophic and autotrophic bacteria. It can also occur under either aerobic or anaerobic conditions. Kawai et al. (1964) found that about half the anaerobic bacteria in a marine system could reduce nitrate.

Denitrification exhibits a fall in free energy when the hydrogen is supplied from an organic source (eq. 4). However, if the reaction proceeds by simple ionization, there is no energy loss (eq. 5). Both reactions, 4 and 5, probably occur in filter beds.

$$4NO_3^- + 3CH_4 \rightleftharpoons 2N_2 + 3CO_2 + 6H_2O \qquad (4)$$
$$\Delta G^\circ = -475 \text{ kcal}$$

$$2NO_3^- + 2H^+ \rightleftharpoons N_2O + 2O_2 + H_2O \qquad (5)$$
$$\Delta G^\circ = +21 \text{ kcal}$$

Denitrification is evident in old culture water when low levels of nitrite chronically persist and the nitrate level decreases.

Mineralization, nitrification, and denitrification are parts of the nitrogen cycle. The mechanisms in nature and in captivity are the same; the effects are not. The natural dispersal of animals in the wild, as a means of overcoming environmental stress, cannot be duplicated within the boundaries of captivity. Captive animals are at the mercy of their limited environment and their lives depend upon the rates of the vital conversions mentioned above.

1.2   CARE OF THE FILTER BED

Salinity

ZoBell and Michener (1938) have questioned whether there are marine and freshwater species of bacteria per se. They found that most marine forms could also be grown in fresh water. Many could survive the change directly from sea water. Moreover, 12 species that superficially appeared to be "marine" bacteria were all successfully converted to fresh water by gradual dilution of their seawater medium in increments of 5 per cent.

Nitrifiers are among those unable to make a sudden transition. Kawai et al. (1965) found that nitrifying activities in a marine system were greatest when the sea water was at normal salinity. Nitrification diminished as the solution was either diluted or concentrated, although some activity remained even after the salinity was doubled.

Nitrification in fresh water was greatest before addition of sodium chloride. At the normal concentration of sea water, nitrification in the freshwater system stopped altogether.

Kuhl and Mann (1962) showed that nitrification proceeded more rapidly in freshwater than in seawater systems, with greater quantities of nitrite and nitrate being produced in sea water. Kawai et al. (1964) also obtained evidence of this, as shown in Fig. 2.

Bacteria in a filter bed can adjust to gradual changes in salinity, but not to sudden fluctuations. When salinity changes rapidly, many bacteria are killed and the metabolism of the survivors is temporarily repressed. There is a time lag while those organisms still living adapt to the new conditions. Meanwhile, ammonia accumulates. In most situations complete adaptation takes several days, which is enough time for the ammonia to reach levels that are toxic to the animals inhabiting the system.

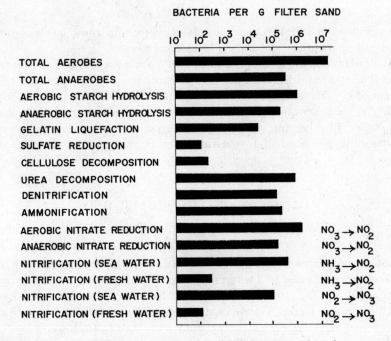

Figure 2. Population of filter bed bacteria in small freshwater and marine systems after 134 days.

In marine systems, surface evaporation causes a gradual increase in salinity. It is not a good practice to allow salinity to increase too much before adding makeup water. If specific gravity is used to measure the density of the medium and a reading of 1.025 is taken as the standard for sea water (see Sect. 6.1), the system should not be allowed to vary more than ±0.002.

Maintaining consistent salinity in brackish-water systems is more difficult. Dilutions of sea water should be made in a separate container and the water added to the system a little at a time. The system should be allowed to circulate completely before more is added. By mixing the makeup water separately, mistakes are relegated to the container and the equilibrium of the system is not jeopardized. Also, diluting the sea water first minimizes the possibility of salinity fluctuations in an already dilute medium.

Aged tap water should be used for makeup in both freshwater and seawater systems. *Aged tap water* is tap water that has been held in an open container for at least 3 days and aerated to expel the chlorine.

Gravel Surfaces

Kawai et al. (1964) determined that nitrifiers in the filter bed were 100 times as plentiful as those suspended in the water. This indicated that an important factor in nitrification was the number of available surfaces for bacterial attachment. The greatest surface area in a culture system is provided by the gravel grains.

There is also evidence that accumulated detritus provides additional surfaces and improves nitrification. According to Saeki (1958), 25 per cent of all nitrification in culture systems is accomplished by bacteria attached to detritus particles. In tests performed by Kawai et al. (1965), 1 g of surface sand from an old marine system was removed and gently washed in clean sea water. Afterward, 40 per cent of the nitrifying ability of the sand was lost. Subsequent washings decreased it even more. When another gram was washed vigorously, 66 per cent of the nitrifying capacity was lost, and this was reduced another 14 per cent by a second washing. These findings showed two things: first, a considerable portion of the total number of nitrifiers is attached to the detritus and, second, vigorous washing detaches others from the gravel surfaces.

Most of the nitrification in a filter bed occurs on the upper gravel layers. Kawai et al. (1965) found that there were 10^5 ammonia oxidizers per gram of sand in the top portions of the filter in a marine system and that the number of nitrite oxidizers was 10^6. At a depth of only 5 cm, the population of each type fell by 90 per cent.

Since most of the biological activity in a filter bed is concentrated in the upper layers, systems should be designed with the surface area of the bed in mind, rather than volume of water. For instance, system A has the following measurements: length, 2 ft; width, 2 ft; and depth, 4 ft. In system B, the length is 4 ft; width, 2 ft; and depth, 2 ft. Both are equipped with subgravel filter plates, and the area of each filter bed is the same as its respective system. Both systems hold identical volumes of water, yet system B supports a greater animal load, because the surface area of its filter bed is twice as large.

Size of the gravel is important. Small gravel has more surfaces for bacterial attachment than comparable weights of large gravel. For example, 6 cubes, each weighing 1 oz, have a total of 36 unit surfaces, whereas 1 cube weighing 6 oz has only 6 surfaces, each larger than the individual surfaces of the small cubes. The total surface area of the 6 1-oz cubes is 3.3 times greater than the surface area of the 6-oz cube.

Circulation through the bed is impaired when the gravel grains are too small. As detritus accumulates and coats the surface of the bed, vertical channels form and the water follows these paths of least resistance. The result is erratic oxygenation and anoxic areas where the growth of aerobic bacteria is inhibited. For this reason sand or very fine gravel is undesirable, especially in deep beds. Gravel measuring 2–5 mm is best for most systems (Saeki, 1958).

The shape of the gravel is also important. Angular gravel has more surfaces than round types. A sphere has the smallest surface area per unit volume of any geometric shape. Coarse, angular gravels are therefore preferable to smooth, water-worn varieties.

The filter bed is a permanent installation. The gravel should never be taken out of the system and washed. Washing removes most of the detritus, which supports a large population of nitrifiers. It also detaches bacteria from gravel surfaces. In cases where it is absolutely necessary to wash the gravel, it should be done directly in the system with clean water of the same salinity. In marine systems, clean sea water should be used; in brackish- and fresh-water systems, clean brackish water and aged tap water, respectively, should be used.

Oxygenation

A filter bed can be compared to a huge, respiring organism. When functioning properly, it consumes a considerable amount of oxygen. The oxygen consumption of microorganisms in a bed is called BOD (Biological Oxygen Demand). In filtration, BOD is measured in terms of Oxygen Consumed during Filtration (OCF).

OCF is partly a function of nitrification. Hirayama (1965) showed that if the BOD of a filter bed was high, a sizeable population of nitrifiers was at work. Hirayama filtered cultured sea water through a column of sand from an old filter bed. Before entering the column, the water had a dissolved oxygen level of 6.48 mg/liter; after passing through 48 cm of sand, it measured 5.26 mg/liter. At the same time, ammonia decreased from 238 to 140 meq/liter, and nitrite from 183 to 112 meq/liter.

Both aerobic and anaerobic bacteria are found in filter beds, but in well-aerated systems aerobic forms predominate. Kawai et al. (1965) demonstrated that nitrification was more efficient when the oxygen tension was high, although some conversion of ammonia and nitrite was noted even at very low oxygen tensions. This was especially true in freshwater systems, in which an actual acceleration in the oxidation of nitrite to nitrate was seen. Marine systems, on the other hand, seemed less efficient under conditions of reduced oxygen tension.

Anaerobic bacteria are inhibited by the presence of oxygen, and adequate circulation through the filter bed holds them in check. When the oxygen tension in the system decreases, however, anaerobes proliferate. Many of their metabolites are toxic.

Both ammonification and deamination take place under anaerobic conditions, but the mechanisms and end products are different. In anaerobic situations, the mechanisms are fermentative rather than oxidative. During ammonification this results in the formation of organic acids instead of bases, along with carbon dioxide and ammonia. These substances, plus hydrogen sulfide, methane, and several others are what give a suffocating filter bed its familiar putrid smell.

The *turnover rate*, or rate at which water moves through a system, should never fall below 1 gsfm (gallons per square foot per minute) in systems of 200 gal or more. This rate keeps the oxygen level near saturation at all times under normal culturing conditions. A turnover rate of 1 gsfm is too rapid in systems holding less than 200 gal. In these systems, the rate should be adjusted to a point where the dissolved oxygen level (see Sect. 9.4) is near saturation with a minimum of surface agitation.

An *airlift pump* is the most trouble-free means of moving water through a biological filter. The advantages of an airlift over mechanical pumps are:

1. Lower initial cost
2. Lower maintenance (an airlift has no moving parts)
3. Easy installation
4. Easily made portable
5. Nonclogging
6. Requires little space
7. Simplicity of design
8. Easy construction
9. Greater efficiency than centrifugal pumps when operating at low head and high submergence
10. Flow rate easily regulated
11. Highly versatile application

An airlift is essentially a vertical length of pipe. In culture applications, part of the pipe extends below the subgravel filter plate and the rest above the filter plate; a portion of it extends above the surface of the water. This is illustrated in Fig. 3.

When a pipe is submerged in water in a vertical position, the levels inside and outside the pipe equilibrate. Air is lighter than water and when air is injected at the lower end of the pipe it forms bubbles and rises. As it rises, it produces a mixture of air and water which is lighter than water alone. The air–water mixture inside the pipe is therefore lighter than the water outside the pipe and the equilibrium is upset. When this happens, heavier water from underneath the filter plate moves into the lower end of the pipe. As long as air is injected, equilibrium never occurs and the air–water mixture is spilled out the top of the pipe.

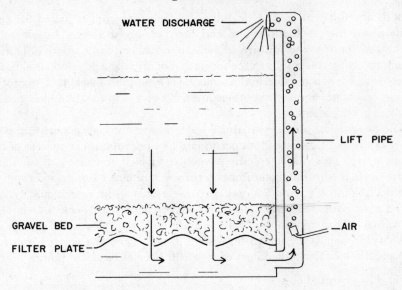

Figure 3. Water circulation through a subgravel filter by the airlift method.

The main factor affecting the efficiency of an airlift is the *per cent submergence* of the lift pipe. The volume of air necessary to operate an airlift increases with decreasing per cent submergence. Maximum efficiency is attained at 100 per cent submergence.

Per cent submergence is a simple calculation. In Fig. 4, if the distance between the air inlet and the discharge point (*A* to *C*) equals 3 ft and the total lift (*A* to *B*) equals 1 ft, then

$$\frac{3 \text{ ft} - 1 \text{ ft}}{3 \text{ ft}} = 66 \text{ per cent submergence}$$

The relationships among per cent submergence, capacity and flow rate, and the diameters of the lift pipes and air lines for airlift pumps with side air-inlets are given in Table 1. As a general rule, doubling the diameter of the lift pipe increases its capacity 5.6 times.

An airlift is less effective when the volume of the injected air exceeds the capacity of the lift pipe. This inefficiency is easily detected by the gurgling sound of air escaping through the top. It can be overcome by decreasing the air flow from the compressor. The effluent water should emerge in a smooth, even stream. If it spurts, the cause can usually be traced to one of two factors: either the air volume is too great for the diameter of the lift pipe and much of the air is escaping directly through the water and into the atmosphere, or the per cent submergence of the lift pipe is not great enough.

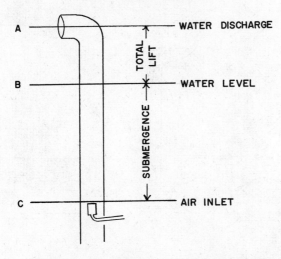

Figure 4. Operating principles of an airlift.

Table 1. Airlift Capacities under Various Conditions Using the Side Air-Inlet Method

Pipe sizes		Capacity, gpm			
Discharge pipe, in.	Air-line sizes	70% submergence	60% submergence	50% submergence	40% submergence
1	$\frac{3}{8}$	10– 17	8– 12	7– 11	6– 10
1$\frac{1}{4}$	$\frac{3}{8}$	16– 24	11– 18	10– 15	8– 12
1$\frac{1}{2}$	$\frac{1}{2}$	20– 36	16– 28	12– 21	10– 19
2	$\frac{3}{4}$	33– 65	26– 55	20– 40	18– 35
2$\frac{1}{2}$	1	60–100	50– 85	36– 60	32– 55
3	1	90–130	78–120	55–100	50– 95
3$\frac{1}{2}$	1	120–250	110–180	90–150	80–130
4	1$\frac{1}{4}$	200–325	160–250	130–200	120–180
4$\frac{1}{2}$	1$\frac{1}{2}$	250–475	200–375	170–275	155–225
5	1$\frac{1}{2}$	300–600	275–475	200–375	180–300
6	2	500–900	450–775	350–575	280–500

Greater efficiency is attained if the air entering the lift pipe is diffused. In small systems, where the diameter of the lift pipe is 1 in. or less, an airstone is sufficient to disperse the air. In larger lift pipes, baffles constructed from PVC (polyvinyl chloride) pipe and containing numerous small perforations should be inserted at the end of the air line at the point of injection inside the lift pipe.

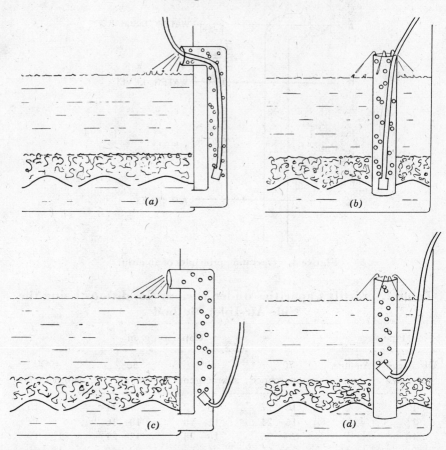

Figure 5. Two central airline designs (*a* and *b*), and two side air-inlet designs (*c* and *d*) commonly used in culturing.

Two basic airlift designs are used in culture applications: the *central air line* (air line inside the lift pipe), and the *side air-inlet* (air line outside the lift pipe). Either is suitable. Two variations of each type are illustrated in Fig. 5.

The *under-* or *subgravel filter* consists of a perforated plate suspending the gravel bed above the bottom of the tank. When used in combination with an airlift pump, no means yet devised can surpass its efficiency for sustaining biological filtration in the filter bed.

Oxygen delivery through the gravel column is efficient and simple with subgravel filtration. Deoxygenated water from animal respiration and OCF is taken from underneath the plate by the airlift and spilled back across the surface of the culture system where it is reoxygenated. This is illustrated in Fig. 3.

An effective subgravel filter plate should cover the entire bottom of the tank and be sealed in all corners and along the edges where it meets the tank walls. Proper sealing accomplishes two things: first, it eliminates dead areas in the gravel where anaerobic bacteria can proliferate and produce noxious substances (filter plates that cover only a portion of the bottom are undesirable for this reason), and second, it prevents gravel from working under the plate and impeding the flow of water to the airlift. The subgravel filter is only effective as long as it forms a false bottom separating the gravel bed from the floor of the tank.

The filter plate must be strong enough not to collapse under the weight of the gravel. Commercial plates are adequate in smaller systems, since most of them are ribbed for added strength. In larger systems, where custom-made plates are necessary, any inert material can be used to support the plate at intervals; for example, bricks coated with fiberglass resin or epoxy paint. The plate itself should be constructed from an inert material. Metals are undesirable. Fiberglass is preferable because it is inert, noncorrosive, strong, and relatively inexpensive. It is also an easy material to work with.

In large systems it is best to make the plate from stock materials to reduce costs. One of the best materials is corrugated fiberglass roofing obtainable at lumber yards and construction supply companies. The sheets are normally 2 ft × 8 ft. The best way to perforate fiberglass roofing is to run it over a table saw, equipped with a special fiberglass-cutting blade, at right angles to the corrugations. The slits should be about 1 in. apart. It is only necessary to slit one side of the plate. The ridges on the other side should be left as they are. Water can pass freely along the sloping sides of the ridges and down through the slits in the troughs, as illustrated in Fig. 3. If a table saw is not available, holes can be drilled in the plate by using a small bit and electric hand drill. The intervals between the holes should be about 1 in.

The perforations in the plate should be smaller than the filter gravel. Otherwise, gravel will work its way underneath.

The plate should fit tightly in the tank to reduce the cost of sealant around the edges. The plate should be placed *slits down* on the supports and checked with a level to be sure that all areas are equidistant from the bottom of the tank.

The most effective sealant is either General Electric or Dow Corning construction grade silicone sealant. This is applied with a caulking gun along the seams between the plate and the tank walls. Fiberglass tape can be used to bridge large gaps between the filter plate and the tank wall. The edges of the tape must be embedded in the sealant.

The plate should set for at least 24 hr after installation before adding gravel and water. Water may hinder the drying of the sealant and the weight of the gravel can break the seal along the edges.

pH

Kawai et al. (1965) found that lowering the pH from 9.0 was more detrimental to nitrification in marine- than in freshwater systems. They attributed this to the normally lower pH values of fresh waters.

Saeki (1958) determined that the conversion of ammonia in freshwater systems was inhibited by low pH. The ideal pH was 7.8. Nitrite conversion took place most rapidly at a pH of 7.1, and he recommended a range of 7.1–7.8 for culturing in fresh water.

In marine systems, the practical pH range for nitrification is 7.5–8.3.

Temperature

Bedford (1933) found that the majority of bacteria in sea water had a temperature growth range of —5 to 37 C. Optimum growth seemed to be in the higher ranges. This condition illustrates the extreme temperature variations bacteria are able to tolerate and still carry on their life functions.

ZoBell (1934) found that nitrification took place in cultured species at —2 C, although the rate of oxidation was greatly reduced. However, after the first few weeks the oxidative processes increased, presumably as the organisms adapted to the cold.

In marine systems, Kawai et al. (1965) found that nitrification was most efficient when the temperature was 30–35 C. In freshwater systems, optimum temperature was 30 C. When the temperatures in both systems were lowered, the inhibitory effects were much less in fresh water.

The optimum temperature selected for a system should be based on the thermal requirements of the animals. The filter bed is not adversely affected by temperature changes. For example, a sudden drop in temperature has no effect on nitrifiers except to temporarily lower their metabolism. The optimum temperatures given above for nitrification are higher than the tolerable limits of most aquatic animals. In the routine care of the filter bed, temperature is the least critical factor, although it is extremely important to the animals under culture (see Sect. 5.2).

Toxic Additives

Many inorganic and organic compounds, including antibiotics, inhibit nitrification (Tomlinson et al. 1966). Formalin interrupts nitrification even at very low concentrations (Burrows and Combs, 1968). When toxic substances are added to a culture system there is a possibility of two inhibiting mechanisms: either the growth and proliferation of the filter organisms are repressed, or the compound has no effect on growth and proliferation, but influences the metabolism of the cells and prevents them from reaching their full oxidizing capacity. Bacteria are highly adaptable and the effects of a given compound on nitrification are difficult to determine. Such measurable effects are further complicated by the chemical complexity of viable culture water.

Nothing should be added to a culture system that upsets the delicate equilibrium of the filter bed. If sick animals are to be treated, remove them from the system and threat them elsewhere. There is no advantage in ridding an animal of its parasites if biological filtration is interrupted in the process. The rise in ammonia following the treatment of culture water with such things as antibiotics or copper is often enough by itself to kill the animals.

The culture area should be completely protected from external contamination. Tobacco smoke and insecticides, for example, are toxic and can enter the water from the air.

1.3 DETERMINATION OF CARRYING CAPACITY

An important aspect of biological filtration is carrying capacity. *Carrying capacity* is defined as the animal load that a system can support. Hirayama (1966) derived the following formula for calculating the carrying capacity of a small marine system[1]:

$$\sum_{i=1}^{p} \frac{10W_i}{\dfrac{0.70}{V_i} + \dfrac{0.95 \times 10^3}{G_i D_i}} \geq \sum_{j=1}^{q} (B_j^{0.544} \times 10^{-2}) + 0.051F$$

The left-hand term represents the oxidizing capacity of the filter bed (OCF) measured as milligrams of O_2 consumed per minute, where W = the surface area of the filter bed (m²), V = filtering velocity (cm/min), D = gravel depth (cm), and p = the number of filters serving the system. In the above formula, G represents the grain-size coefficient of the gravel grains. This is determined by

$$\frac{1}{R_1} x_1 + \frac{1}{R_2} x_2 + \frac{1}{R_3} x_3 + \ldots + \frac{1}{R_n} x_n$$

where R = the mean grain size of each fraction of gravel in the bed (if the gravel is graded) in millimeters, and x = the percentage weight of each fraction.

---

[1] The same relationship may also hold true for freshwater systems, but this needs verification.

The right-hand term of the original equation represents the rate of "pollution" by the animals. It is also expressed as milligrams per minute of O_2. In this term, B = the body weights of the individual fishes (in grams), F = the amount of food (in grams) entering the system daily, and q = the number of fishes in the system.

As seen from the formula, the oxidizing capacity of the filter bed must be greater than, or equal to, the rate of pollution by the animals. It is also important to note that *as the weights of the individual animals decrease, the carrying capacity of the system decreases.* In other words, carrying capacity is not merely a function of the total animal weight. A system that can support a single 10-lb fish cannot necessarily support 10 fishes each weighing 1 lb. Let us assume, for example, in a hypothetical system with one filter that $W = 0.35 \text{ m}^2$, $V = 10.5 \text{ cm/min}$, and $D = 36 \text{ cm}$. If the gravel is all the same grade and $R = 4 \text{ mm}$, then $G = \frac{1}{4} \times 100$, or 25.

Substitution of these values into the left-hand term of the original equation gives the OCF value:

$$\frac{10\,(0.36)}{\dfrac{0.70}{10.5} + \dfrac{0.95 \times 10^3}{25\,(36)}} = \frac{3.6}{\dfrac{0.067 + 950}{900}} =$$

$$\frac{3.6}{0.067 + 1.055} = \frac{3.6}{1.122} = 3.2 \text{ mg/min OCF}$$

Let us further assume that a single fish is to be cultured in the system and that the fish will not be fed. Looking at the right-hand term, if X represents OCF, then

$$X = \sum_{j=1}^{q} (B_j{}^{0.544} \times 10^{-2}) + 0.051F$$

In this particular case, $q = 1$ and F does not enter into the equation, since the fish is not being fed. Therefore, the largest weight of a single fish, B, that the system will support is determined by

$$3.2 = B^{0.544} \times 10^{-2}$$

$$B \cong 40,700 \text{ g}$$

or approximately one 90-lb fish. Now suppose that only fishes weighing 1 lb each are to be cultured in the same system. They will not be fed either. The largest number of 1-lb fishes, q, that the same system would support is

$$3.2 = q\,(454^{0.544} \times 10^{-2})$$

$$q \cong 11.5$$

or approximately 11 1-lb fishes. Daily feeding reduces the carrying capacity even more. Assuming that each 1-lb fish is fed 6 g of food daily, the largest number of fishes that could be supported is

$$3.2 = q (454^{0.544} \times 10^{-2}) + 0.051 (6q)$$

$$q \cong 5.46$$

or approximately 5 1-lb fishes.

1.4 THE "CONDITIONED" SYSTEM

A *conditioned* system is one in which the filter bacteria are in dynamic equilibrium with the routine formation of their energy sources.[2]

Nitrification can be used as a yardstick for determining when a new system becomes conditioned and suitable for culturing. At first a high ammonia level is the principal limiting factor. Ammonia usually subsides within 2 weeks under warm-water conditions (ambient temperature 15 C and up) and after a slightly longer period in cold water (ambient temperature below 15 C).

Although a system may be ready for culturing after 2 weeks, it is not completely conditioned because many of the important groups of bacteria have not yet stabilized. Kawai et al. (1964) described the bacterial population in a warm-water marine system after 3 months as follows. (The results are shown graphically in Fig. 2.)

1. *Total aerobic.* Increased tenfold within 2 weeks after the addition of fishes to the system. Maximum population density of 10^8 organisms per gram of filter sand was reached after 2 weeks. After 3 months the population stabilized at 10^7 per gram of filtrant.

2. *Protein decomposing.* Original population started at 10^3 per gram of filter sand and increased 100-fold after 4 weeks. The population stabilized at 10^4 after 3 months. The reason for the dramatic increase was thought to be due to the food (fish flesh), which is high in protein.

3. *Starch decomposing.* Original population was 10 per cent of the total number of bacteria. There was a gradual increase, then a decrease after 4 weeks. The population stabilized at 1 per cent of the total population after 3 months.

4. *Nitrifiers.* Maximum density of nitrite formers was reached after 4 weeks and that of nitrate formers after 8 weeks. Nitrite formers were present in

---

[2] The term "conditioned," as used here, is not to be confused with its older interpretation in which the culture water was not aerated, yet the longevity of the animals notably increased, presumably as the water aged and developed bacteriostatic properties (Atz, 1964b).

greater numbers than nitrate formers after 2 weeks. Stabilization occurred at 10^5 and 10^6 per gram of filter sand, respectively.

In the initial stages of nitrification there is a time lag between the fall of ammonia and the oxidation of nitrite. This is because the growth of *Nitrobacter* is inhibited by the presence of ammonia (Lees, 1952). Efficient oxidation of nitrite does not take place until most of the ammonia has been converted by *Nitrosomonas*.

The high ammonia level seen in a new system results from an initial population imbalance between heterotrophic and autotrophic bacteria. When a new system is put into operation, the growth of heterotrophic species at first exceeds the growth of autotrophs. Much of the ammonia produced from ammonification and deamination is used by part of the heterotrophic population. In other words, there is no sharp dividing line between heterotrophic and autotrophic bacteria with respect to ammonia utilization, since many heterotrophic species also use it as an energy source. Ammonia oxidation by nitrifiers proceeds only after the heterotrophic population has subsided and stabilized (Quastel and Scholefield, 1951).

The number of bacteria in a new system is a significant factor only until each type has become stabilized. Subsequently, increases in the metabolic activities of the individual cells compensate for fluctuations in energy sources, but there is no additional increase in the number of cells. Work by Quastel and Scholefield (1951) has shown that the population densities of nitrifying bacteria occupying a given number of surfaces are relatively constant and independent of the quantity of the available energy source.

The total oxidizing capacity of the bacterial population in a conditioned system is geared to a stable daily input of oxidizable substrates. Sudden increases in the number of animals, their weights, and the quantity of food added each day often result in measurable rises in the ammonia and nitrite levels. These levels may persist until the bacteria equilibrate with the new conditions.

The extent of the ammonia and nitrite increases depends on how much the additional load has stressed the carrying capacity of the system. If the increase in the animal and food load is still beneath the maximum carrying capacity, then equilibrium with the new conditions is usually attained within 3 days in warm-water systems and after a slightly longer time in cold water. If the additional load pushes the system beyond its maximum carrying capacity, the results are permanently increased ammonia and nitrite levels.

Ammonification, deamination, nitrification, and denitrification are processes that follow one another more or less in sequence in a new system. Once the system is conditioned, they all occur simultaneously. In a conditioned system, the measurable ammonia (as total NH_4^+) is less than 0.1 ppm, and measurable nitrite is the result of denitrification. These levels remain constant and there are no time lags in the conversion processes because all energy sources are oxidized simultaneously.

It is best to *overcompensate* a system when conditioning it; that is, to condition it with a slightly greater animal load than it will ultimately carry. Overcompensation eliminates these later increases in ammonia entirely.

Only hardy animals should be used to condition a system. Animals highly susceptible to ammonia poisoning should not be added until nitrification is fully established. Turtles are excellent to condition new systems. They are less affected by ammonia than fishes and invertebrates and yet they continuously supply the organics necessary to initiate biological filtration. Marine turtles and diamondback terrapins are suitable for conditioning seawater systems, and any of the common freshwater forms, such as sliders, snappers, and map turtles, are ideal for fresh water. Among fishes, moray eels, groupers, carp, and many of the catfishes are notably tolerant of ammonia.

Adding specimens a few at a time is always a good technique. If no hardy animals are available and if the species to be cultured are sensitive to ammonia poisoning, the animal load can gradually be built up to maximum density. For instance, if it is necessary to keep the ammonia level below 0.2 ppm at all times, then the animal population can slowly be increased at a rate which will not exceed this level. The technique is to monitor the ammonia continually and not add specimens faster than the nitrifying population can stabilize at 0.2 ppm or less. This technique involves much laboratory work and prolongs the conditioning time. The first method—overcompensating with hardy animals—is quicker and more practical.

The time lags in cold water are longer because the growth of bacteria is slower at low temperatures. The process can be speeded up by maintaining the system at warm-water temperatures with warm-water animals until nitrification is established. The warm-water animals can then be removed, the temperature dropped, and a similar (preferably lower) weight of cold-water specimens added. It does not matter how fast the temperature is decreased after the warm-water animals are taken out. However, upon adding the cold-water animals, slight increases in ammonia and nitrite are sometimes noted, even if the system has been initially overcompensated. These usually subside after 3 days, indicating that the bacteria have adapted to the cold. Such increases can be minimized by allowing 48 hr for the bacteria to adjust to the lower temperature before adding the cold-water animals.

There is only one reliable method to accelerate the conditioning process. A portion of the top gravel from a conditioned system can be added to the surface of the new filter bed. Part of the detritus should be included, since it contains substantial numbers of bacteria.

1.5   MECHANISMS OF DETRITUS FORMATION

The function of detritus in nitrification has already been mentioned. In closed-system culturing, *detritus* is defined as the loosely aggregated material that accumulates in culture systems, usually on the gravel in the filter bed, but also on stationary objects in the system when circulation is poor. Biochemically, its composition is extremely complex and as yet undefined. It consists of both organic and inorganic substances, and heterotrophic bacteria are important in its formation. Detritus formation is an endless process as long as the water in a culture system contains living organisms and dissolved organic matter. Methods for its removal are discussed in the next chapter.

Air bubbles in the water are significant in detritus formation. Baylor and Sutcliffe (1963) demonstrated that organic aggregates were formed on bubbles by the adsorption of organic material from the medium. Riley (1963) and Sutcliffe et al. (1963) caused aggregates to form by bubbling air through microfiltered sea water. Aggregate formation indicated the extraction of organic substrates from solution.

Detritus aggregates continue to grow in size from the time of their initial formation. Two mechanisms are responsible: agglutination and further adsorption (Riley, 1963). In culture systems, agglutination is probably the dominant factor at the surface of filter beds where particulate matter is concentrated. Aggregate particles adhere to gravel grains at first. As their numbers increase, they fill the interstices among the gravel grains and also adhere to each other by electrostatic attraction. Eventually they grow to clearly visible size.

The second factor responsible for the growth of detritus aggregates—additional adsorption—is significant at the air–water interface, particularly at the point of water discharge into a system where surface agitation is heaviest. It may also be important inside airlifts where the dispersion of air into fine bubbles increases the number of surfaces on the bubbles for organic adsorption. According to Barber (1966), aggregates are formed at the surface of the water under natural conditions

. . . in a physical process by which organic molecules come out of solution to form an organic skin around a rising bubble. At the surface of the sea the bubble is ejected into the air, leaving behind its organic skin as a collapsed monomolecular film which may then sink through the water column, providing a seed on to which additional dissolved material may aggregate.

A similar process probably takes place in closed systems. Carlucci and Williams (1965) demonstrated that bubbling air into sea water produced concentrated numbers of bacteria in the resulting froth. It is probable that the froth collected from airstripping (see Sect. 3.3) also contains large numbers of bacteria, except when ozone is used. Newly formed aggregates in the water column quickly acquire a population of bacteria. The aggregates represent a consolidated energy source and also additional surfaces for attachment, especially in the heavily populated areas of a system, such as the filter bed. This accounts for the findings by Kawai et al. (1965), indicating a severe decline in the nitrifying ability of filter sand after the detritus had been washed away.

Sieburth (1965) noted that even without bubbling, natural populations of bacteria in sea water formed detritus during deamination and ammonification. He offered a possible mechanism for the combination of organic and inorganic substances during detritus formation. According to Barber (1966), Sieburth

. . . suggested that the microzonal alkalinization that must occur during ammonia liberation could initiate the precipitation of inorganic nuclei around which organic aggregates could form.

Detritus is essentially harmless, although when it accumulates in dense mats other problems arise, as will be shown in the next chapter. The aggregates are no longer in solution and are therefore less available for adsorption by animals. The toxic properties of a dissolved substance are greatly reduced as soon as it is extracted from solution and deposited in the detritus. This principle, in fact, is used in chemical filtration with activated carbon (see Sect. 3.2); the substance adsorbed in the pores of the carbon is technically no longer in solution and therefore not available for secondary adsorption by the animals.

The mechanisms of detritus formation offer possibilities as feeding techniques for microscopic larvae. Baylor and Sutcliffe (1963) demonstrated that *Artemia* could be raised on aggregate material that was formed after bubbling air through sea water. Their discovery has interesting potential. For instance, a nutrient solution could be poured into an air column prior to feeding a batch of larvae. The nutrients would partially aggregate in the column and become available to the animals. After a period of time in which the nutrient water was cycled past the animals, normal water flow could be resumed and the excess material removed by chemical filtration.

CHAPTER 2

Mechanical Filtration

2.1 DEFINITION AND FUNCTION

Mechanical filtration is the physical separation and concentration of suspended particulate matter from circulating water. It is accomplished by passing the water through suitable substrata or septa that trap the particles. The trapped material is then removed by various methods, depending on the type and design of the filter.

The functions of mechanical filtration in closed-system culturing are three-fold: (1) to reduce the turbidity in the water caused by suspended microorganisms and other particulate matter, (2) to lower the level of organic colloids, and (3) to remove accumulated detritus from biological filter beds. Mechanical filters are also used for prefiltering natural water, a procedure that reduces turbidity and removes large numbers of microorganisms that might dangerously increase the BOD of the systems or cause epizootic outbreaks of disease.

2.2 MECHANISMS OF MECHANICAL FILTRATION BY GRAVEL

Gravel in biological and rapid sand filters (see Sect. 2.4) reduces turbidity in water by trapping particulate matter and removing it from suspension. This is accomplished in two ways. First, suspended matter is physically trapped in the interstices among gravel grains. Second, the electrostatically charged surfaces of the gravel grains attract oppositely charged particles or colloids and remove them from solution. The efficiency of these processes depends upon the factors discussed below.

Size of Gravel

The mechanical filtering efficiency of gravel increases with decreasing grain size of the individual granules. Smaller granules have more surfaces exposed to the water for electrostatic attraction of particulate and colloidal matter. Also, the smaller interstices facilitate removal of finer particles. This results in a greater percentage of suspended matter removed per volume of filtered water.

Accumulation of Detritus

Accumulated detritus reduces the size of the interstices among gravel grains, enabling the filter bed to remove finer particles. This is the main reason why old filter beds give better clarity.

Shape of Gravel

Rough, angular gravels are best for mechanical filtration. Their numerous surfaces increase the electrostatic potential of the bed. Irregular granules also reduce the number of consecutive interstices through the gravel column. In biological filters, this keeps detritus from working into the deeper layers of the bed where it is more difficult to remove by the usual cleaning methods.

Grading

Only one grade of gravel is used in airlifted biological filters not equipped with backwashing mechanisms. When different grades are mixed, the number of surfaces is reduced. Large voids form in the gravel column in the areas where larger grains predominate. Detritus can work deeper into the bed in these localized "loose" regions, and in shallow beds it can circulate completely through and back into the culture system.

Gravel in rapid sand filters is purposely graded so that filtration can also take place in the deeper layers.

Even Distribution of Gravel

The path that water takes in its downward movement through the gravel depends upon localized patterns of resistance within the bed. Water flow in biological filters is distorted when gravel is unevenly distributed on the filter plate. Thin parts of the bed offer less resistance than thick parts and attract a greater amount of circulating water. This condition may lead to the same chronic turbidity problems often seen in systems with ungraded beds.

Even distribution of each grade of gravel is also important when initially filling a rapid sand filter. This assures well distributed circulation through the gravel column. In rapid sand filters it is necessary for the uniform expansion and settling of each graded layer during backwashing.

2.3   THE BIOLOGICAL BED AS A MECHANICAL FILTER

Factors Affecting Efficiency

The biological filter bed is the only source of mechanical filtration needed in most culture systems. Biological filtering configurations were given in Chapter 1 and they are essentially the same for mechanical filtration.

1. Surface area of the bed equal to the surface area of the culture system.
2. Gravel size = 2–5 mm.
3. Gravel spread evenly over the filter plate.
4. Gravel shape irregular and angular.
5. Turnover rate = 1 gsfm in systems of 200 gal or more.

Most of the suspended particulate matter is trapped in the upper layers of gravel in a standard biological filter, and the surface area of the bed is therefore very important. A standard rule is to make the surface area of the filter at least equal to the surface area of the culture system.

The gravel size (2–5 mm) recommended for biological filtration also works well as a mechanical filtrant. Detritus eventually reduces the gravel interstices, making a top layer of finer gravel unnecessary.

Gravel must be spread evenly over the filter plate, as previously mentioned. This is particularly important in shallow beds (less than 4 in.). There are no firm rules to follow with respect to water volume versus gravel depth. A bed only 2 in. deep may work well in one instance, while 6 in. may be needed in another system of similar dimensions. However, no gravel bed should be less than 2 in. no matter how small the culture system. Systems less than 10,000 gal rarely need gravel more than 48 in. deep if they are well designed and managed.

Irregular, angular gravel is necessary in mechanical filters to trap suspended particulate matter in the upper portions of the bed. Cleaning the bed is easier when detritus has not worked down below the top 6 in.

The turnover rate—1 gsfm in systems larger than 200 gal—is the same as for biological filtration. By sanitary engineering standards, a biological filter operated at this rate is classified as a "slow sand filter." Much of its efficiency depends upon microbial oxidation in the upper regions of the bed and upon entrapment of additional material by surface detritus. Little head loss occurs in this type of filter because bacterial activity maintains the porosity of the gravel column.[1]

---

[1] Throughout this book the terms biological filter, standard biological filter, and filter bed refer to airlifted slow sand filters. This is to distinguish them from rapid sand filters which also have filter beds and can sustain biological filtration.

Cleaning

The complete removal of accumulated detritus is never the purpose in cleaning a biological filter; some detritus must be left to aid in biological and mechanical filtration. However, heavy surface mats (*Schmutzdecke*) are undesirable and must be periodically removed. A Schmutzdecke increases the BOD in a system and reduces its carrying capacity.

Filter beds that have not been cleaned for a long time shrink and pull away from the walls of the container (Fair and Geyer, 1958). When shrinkage occurs, head loss is less near the walls, and channels form along the edges of the bed. The accelerated water flow through these areas causes the heavy deposition of detritus that is seen in corners and along the walls of old filter beds.

Since most of the detritus accumulates in the surface layers in biological filters, periodically stirring the surface of the bed puts most of it into suspension. All airlifts should be shut off while the bed is stirred, because this helps to prevent the suspended detritus from being pulled into the gravel again. Small filter beds can be stirred by hand; garden rakes are handy for larger beds. The gravel in a biological filter bed must be stirred gently. Vigorous stirring detaches significant numbers of nitrifying bacteria from the gravel surfaces. Once the detritus is in suspension, it can be reduced by either siphoning out a portion of the turbid water or by filtering the water with auxiliary devices, such as sand-pressure filters. Either method prevents the formation of a Schmutzdecke, but still leaves enough detritus to maintain biological and mechanical filtration.

When any of the following conditions appear the filter should be cleaned:

1. Formation of a Schmutzdecke.
2. Heavy concentrations of detritus in the corners and along the walls of the filter.
3. Reduced flow rate.
4. Dissolved oxygen below saturation in the filter effluent.

2.4 RAPID SAND FILTERS

Working Principles

Rapid sand filters are powered by mechanical pumps and have turnover rates several times as fast as airlifted biological filters. The degree of clarity is no greater than can be attained with properly managed airlifted filters, although the more rapid turnover may reduce the turbidity in less time.

Neither rapid nor slow sand filters consistently remove particles smaller than 30 μ.

The design and operation of rapid sand filters differ considerably from the design and operation of standard biological filters. The surface area of the bed is not as critical because of the high turnover. Moreover, removal of suspended particulate matter takes place through substantial depths of the bed, instead of only at the surface.

Rapid sand filters are cleaned mechanically by backwashing. This removes more detritus (especially from deep layers) than methods given for cleaning a biological filter. Accumulated detritus is not as critical an efficiency factor, however, since the fine surface sand has small interstices, making detritus less important as a mechanical aid.

Five grades of silica gravel are commonly used in rapid sand filters, including the top layer of sand. These are arranged in horizontal layers of increasing grain size from cobblestones at the bottom of the bed to fine sand at the top. In recent years, anthracite coal has been used as a surface filtrant with sand underneath, followed by three grades of gravel of increasing size. The low density of anthracite helps achieve deeper penetration of suspended particles, offsetting the relatively low surface area of the filter and resulting in greater efficiency.

Microbial oxidation is not as significant in detritus breakdown as it is in standard biological filters. Part of the reason for this is that rapid sand filters must be cleaned more often. The population of bacteria is thus greatly reduced by the elimination of most of the detritus.

Sand-Vacuum Filters

Sand-vacuum filters are open, rapid sand filters in which circulation through the gravel is sustained by removing water from underneath the bed. This creates a partial vacuum at the bottom, and the flow of water from the top is increased. The principle of water circulation is much the same as in standard biological filters, except that mechanical pumps are used and the turnover is higher. This type of filter is commonly used in municipal water supply and waste-water filtration. Most of the standard designs given in waste-water engineering texts can be adapted to large culture installations.

Sand-vacuum filters are recommended for culture systems with volumes greater than 10,000 gal or where large volumes of natural water are prefiltered daily. In both cases, the advantages include the mechanized backwashing arrangements which are more thorough and time saving than stirring by hand.

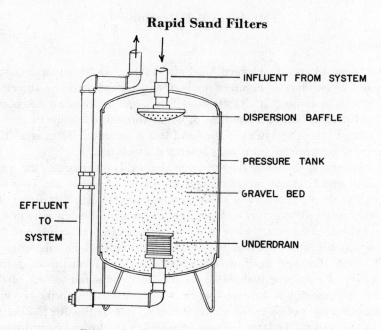

Figure 6. Cutaway view of a sand pressure filter.

Sand-Pressure Filters

The circulation in a sand-pressure filter is illustrated in Fig. 6. Water enters a pressure tank containing the gravel bed through a baffle at the top. Inside the tank the water is forced, under pressure, down through the gravel column.

Water-Saving Techniques Using Rapid Sand Filters

Rapid sand filters are ideal for the intermittent cleaning of airlifted biological filters. In large installations away from the sea, the expense of artificial sea water often prohibits detritus removal by stirring the biological bed and then discarding the turbid water. Also, when warm-water animals are cultured indoors in temperate climates, natural water at winter temperature may be too cold to use directly for makeup. An ideal arrangement utilizes a standard biological filter with an attached auxiliary sand-pressure filter. The influent to the auxiliary unit is installed between the surface of the biological filter bed and the surface of the water. The gravel in the biological filter is stirred and the sand-pressure filter is then turned on. Suspended detritus is pulled into the sand-pressure filter and the clean water is returned to the culture system.

Cleaning

Rapid sand filters are cleaned by backwashing. Detritus accumulates throughout a considerable depth of the bed, and surface stirring is therefore not adequate to remove it. Rapid sand filters are backwashed when the buildup of detritus impedes the flow of water through the gravel bed and causes a head loss. The extent of the head loss is indicated by gauges. The time interval between backwashings is termed a *cycle run*.

In backwashing, the water flow through the bed is reversed. As the wash water rises, gravel grains are lifted and the entire bed expands. Detritus, being lighter than the top layer of sand, rises higher and is forced out the overflow to waste. The gravel and sand grains are kept momentarily in suspension because the upward force of the water and the settling force of the grains are equal. After backwashing, the gravel settles back into the original graded layers. Trapped detritus is removed by three mechanisms during backwashing. First, loose material is simply lifted from the surface of the filter bed and passed out the overflow. Second, detritus deep in the bed and material adhering loosely to the gravel is flushed out by the direct scouring action of the water. The larger grades of gravel in the lower layers disperse the backwash water evenly through the upper layers. Third, a portion of the detritus clinging to the gravel grains is abraded off as the grains are lifted and collide. All three mechanisms inhibit biological filtration. The efficient removal of surface detritus also removes many potential attachment sites for bacteria. The scouring action of the water and the abrasion caused when gravel grains collide remove attached bacteria from gravel surfaces and flush them out with the wash water.

2.5 DIATOMACEOUS EARTH FILTERS

Working Principles

In diatomaceous earth (DE) filtration, a layer of graded skeletal remains of diatoms, held against a porous sleeve by vacuum or pressure, is used to remove particulate matter from water. Because of the expense and maintenance time, this type of filtration is only practical for systems containing at least 1000 gal of water in which a high degree of clarity is required, and organic colloids and suspended microorganisms must be kept at minimal levels. DE filters are capable of removing smaller particles than gravel filters. Some of the finer grades of DE consistently remove particles as small as 0.1 μ.

The filtering mechanism of a DE filter consists of two parts. The first is a porous *central core*, usually made of rigid polypropylene. A series of these are attached to a manifold. The second part is the *filter sleeve*. The sleeve is preferably a thin, tightly woven polypropylene cloth that fits over the central core. Other materials, such as nylon, are less durable. In larger filter units the sleeves are removable. The central core and the sleeve together make up the *filter element*. The sleeve supports a layer of diatomaceous earth; the central core supports the sleeve and exposes it to the circulating water. Figure 7 shows an arrangement with "leaf" filter elements and plumbing. The central cores cannot be seen because they are covered with sleeves. Figure 8 shows an arrangement with "column" elements and manifold.

The layer of diatomaceous earth protects the sleeve from becoming clogged with colloids and losing its porosity. This layer, called a *filter cake*, actually does the filtering. The principle is illustrated diagrammatically in Fig. 9. The figure shows a cross-sectional view through a sleeve and filter cake. Water passes through the filter cake and element, leaving suspended particles and colloids trapped in the cake.

DE Vacuum Filters

Figures 7 and 8 show typical DE vacuum filters. Filtering is done in an open tank, or *bay*. Water enters the bay and passes through the filter elements, the manifold, and into the pump. The principles of water flow are identical to those in sand-vacuum filtration.

DE Pressure Filters

Figure 10 is an illustration of a DE pressure filter. The filter elements are sealed inside a pressure tank.

2.6 FACTORS AFFECTING THE EFFICIENCY OF DE FILTERS

Precoat

After a DE filter is cleaned, it is coated with a new layer of diatomaceous earth, called the *precoat* (Fig. 9). Previous to this, the old filter cake is flushed to waste along with the accumulated particulate and colloidal matter. A new cake must be started to initially protect the sleeves from organic coating. First the water in the filter must be isolated from the water in the culture system—by an arrangement of valves—and recycled as a separate closed system. To precoat, DE is added and the recycling continues until the precoat becomes attached to the elements. Water containing newly added DE

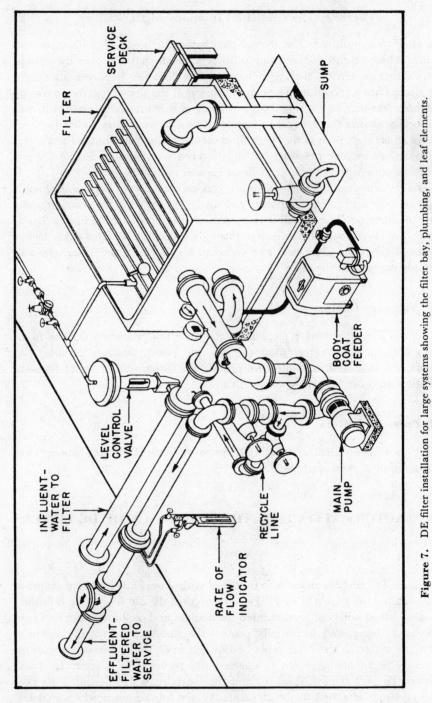

SERVICE DECK

FILTER

SUMP

INFLUENT—WATER TO FILTER

LEVEL CONTROL VALVE

BODY COAT FEEDER

RATE OF FLOW INDICATOR

RECYCLE LINE

MAIN PUMP

EFFLUENT—FILTERED WATER TO SERVICE

Figure 7. DE filter installation for large systems showing the filter bay, plumbing, and leaf elements.

30

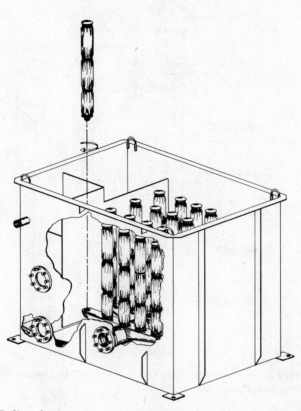

Figure 8. DE filter for large systems showing the filter bay, manifold, and column elements.

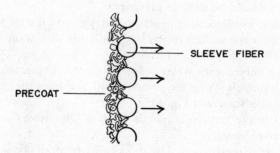

Figure 9. Diagrammatic cross section (magnified) showing the sleeve fibers and beginning of a filter cake, or precoat.

31

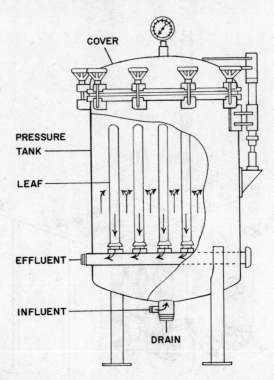

Figure 10. DE pressure filter for large systems showing the pressure tank, gauge, elements, and direction of water flow.

is milky. Precoating is completed when the water turns clear. Normal filtration is resumed by diverting the recycling water back into the culture system. New culture water should be used to precoat.

No special precoat equipment is necessary for vacuum filters. Dry DE is merely sprinkled onto the surface of the bay while the unit is on "recycle." Permanently installed DE pressure filters require a *precoat pot*. The correct weight of dry DE is mixed with water to form a slurry. This is added to the precoat pot, which funnels it into the pressure tank and onto the elements. A typical arrangement is shown in Fig. 11.

The correct amount of precoat is usually 15 lb of DE (by dry weight) per 100 ft² of filter surface area.

The plumbing in all permanent DE units must be designed so that the water flow does not have to be shut off after precoating in order to begin filtering the culture water. In vacuum filters, DE is held against the elements by a partial vacuum inside the central core and in pressure units it is held by pressure applied directly to the filter cake. In both cases, the filter cake starts to fall off the elements as soon as the pumps stop.

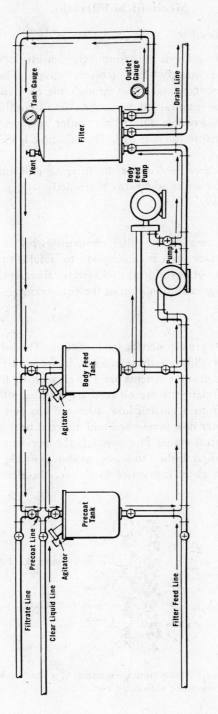

Figure 11. Typical DE filter installation for large systems showing the precoat pot, plumbing, pressure filter, and body feeder.

Body (= Slurry) Feed

The precoat is not sufficient to maintain the porosity of the filter cake because organic coatings quickly build up on the surface. These substances are compressible and restrict water flow through the elements. In permanent units cake porosity is maintained by adding small continuous amounts of comparatively incompressible DE to the outer surface. These measured amounts of DE constitute the *body feed*. The principle is illustrated diagrammatically in Fig. 12.

The body feed largely determines the filtering efficiency of a DE unit. Without it, cycle runs are considerably shortened.

Surface Area

The total surface area of the filter elements is an important factor in efficiency. If the surface area is inadequate to handle the particulate and colloidal matter routinely produced in a system, short cycle runs will result regardless of the design and handling of the equipment.

Cleaning

Most pressure units are cleaned by backwashing. The head loss, indicating restricted water flow through the elements, is shown by pressure gauges. Vacuum units with column elements are also cleaned by backwashing. Vacuum units with leaf elements are cleaned by washing off the cake with a strong jet of water from a garden hose after the bay has been drained. In backwashing, the water flow is reversed and directed back into the manifolds and through the central cores. This expands the sleeves and washes off the filter cake, which is then flushed to waste, as shown in Fig. 13. Backwashing should be done with clean tap water to protect the sleeves from colloidal coating.

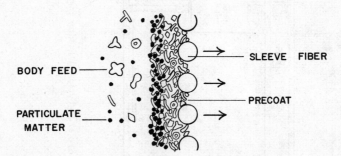

Figure 12. Diagrammatic cross section (magnified) of a filter sleeve showing the sleeve fibers and cake porosity sustained with body feed.

Figure 13. Diagrammatic cross section (magnified) of a filter sleeve showing the sleeve fibers and cake removal by backwashing.

The sleeves should be periodically removed from the central cores and laundered to remove colloids. This can be done in a washing machine with a mild detergent and water softener. Polypropylene sleeves are sensitive to high temperature and should never be washed in hot water. Excessive heat melts the material and closes the interstices in the cloth, and therefore only cold or lukewarm water should be used. After the sleeves are laundered, they should be rinsed several times in clean tap water to remove all traces of detergent. Once on the cores, it is a good idea to backwash the elements several times in succession as an added precaution.

2.7 TROUBLE-SHOOTING CLOGGED FILTER SLEEVES

Filter sleeves eventually become clogged. The signs of clogging are: (1) shortened cycle runs; (2) a vacuum above zero indicated on the vacuum gauge (vacuum filters), or a pressure above zero indicated on the pressure gauge (pressure filters) immediately after backwashing and return to normal filtration; and (3) bare spots on the sleeves after precoating.

Clogging is caused by the following factors: buildup of organic colloids on the sleeves, iron oxide, carbonate scale, algae growth, and manganese. Colloidal clogging is the most common cause of short cycle runs. The problems of iron oxide, carbonate scale, and manganese are more often seen in seawater systems, in which the quantities of inorganic solutes are much higher and the pH is commonly above 8. Algae may present a problem in either fresh or sea water if the water is exposed to direct sunlight. Algae accumulation is more troublesome in semiclosed and open systems in which microscopic organisms are present in the natural water supply.

To determine the cause of clogging, the following key can be used, after the bay has been drained and the elements cleaned.

Troubleshooting Key

(*1*) Fill a pipet with orthotolidine or muriatic acid and squirt it on a clogged spot on an element. Let it stand for 5 min, then rinse with tap water.

   a. The clogging substance does not change color, but the cloth turns white........................CA OR MG CARBONATE. Use *treatment 1*

   b. The clogging substance turns red; after rinsing, the cloth is white ....IRON OXIDE. Use *treatment 1*

   c. The clogging substance turns dark gray or black.............(*2*)

   d. None of the above..(*3*)

(*2*) Dissolve several crystals of sodium sulfite in a few milliliters of orthotolidine or muriatic acid and squirt it on the gray spot. The gray disappears and the cloth turns white.................MANGANESE, Use *treatment 2*

(*3*) *a.* The elements feel greasy.................................(*4*)

   b. The elements do not feel greasy. Squirt a few drops of 25 per cent sulfuric acid on the spot. The spot is white after rinsing............ ....LIGHT ORGANIC COATING. Use *treatment 3*

(*4*) If no positive results are attained at this point, then the elements have a HEAVY ORGANIC COATING, as indicated by the greasy quality of the cloth. This is most often caused by bacterial decomposition of the fatty components in fish flesh and oil (if fish flesh is used as food) and in animal excreta, although it is sometimes caused by algae. If the problem is recurrent, it can usually be traced to one of four conditions: improper maintenance of the filters (e.g., careless backwashing which leaves the elements dirty, or failure to use *treatment 3* routinely every 3 months), inadequate prefiltration with sand, insufficient chemical filtration, or backwashing or precoating with old culture water...Use *treatment 4*

The following treatments should be used in conjunction with the above key. Before starting treatment, the filter should be drained and the elements thoroughly cleaned and inspected. Column elements should be backwashed twice. Both column and leaf elements should be hosed off with a strong jet of tap water from a garden hose (vacuum filters). It is also necessary to know the exact volume of the filter. The chemical additives used in the various treatments are given per 30 gal of water.

Treatment 1

1. Fill the filter with clean tap water. Do not precoat.

2. Be sure the return valve to the culture system is closed.

3. Add 1 qt of 34 per cent muriatic acid per 30 gal of water in the filter to make a 0.25 per cent solution.

4. Recycle for 15 min, or until the cloth turns white. When a heavy concentration of iron oxide is present, the solution will turn brown.

5. Drain the filter and backwash the elements (or hose off leaf elements) three times in succession, letting the unit drain completely after each backwash or hosing.

6. Fill the filter with clean fresh or sea water, precoat, and resume normal filtration. The vacuum or pressure gauge should indicate zero after precoating. No bare spots should be visible on the elements.

Treatment 2

1. Fill the filter with clean tap water. Do not precoat.
2. Be sure the return valve to the culture system is closed.
3. Add the same quantity of muriatic acid as in treatment 1.
4. Gradually add $\frac{1}{3}$ oz sodium bisulfite per 30 gal of water in the unit.
5. Recycle until the cloth turns white.
6. Drain the unit and backwash (or hose) the elements three times in succession, letting the filter drain completely after each cleaning.
7. Fill the filter with clean fresh or sea water, precoat, and resume normal filtration. The vacuum or pressure gauge should indicate zero after precoating. No bare spots should be visible on the elements.

Caution: Do not use this treatment without adequate ventilation.

Treatment 3

1. Fill the filter with clean tap water. Do not precoat.
2. Be sure the return valve to the culture system is closed.
3. Adjust the pH of the water to 5 with muriatic acid.
4. Add 0.1 gal of 15 per cent sodium hypochlorite per 30 gal of water in the unit.
5. Recycle for 3 hr.
6. Reduce after 3 hr by adding 1 oz of sodium thiosulfate per 30 gal of water.
7. Continue to recycle until the chlorine residual (measured by the OTO test) is 0 ppm for three consecutive tests spaced 15 min apart.
8. Drain the unit and backwash (or hose) the elements three times in succession, letting the filter drain completely after each cleaning.
9. Fill the filter with clean fresh or sea water, precoat, and resume normal filtration. The vacuum or pressure gauge should indicate zero after precoating. No bare spots should be visible on the elements.

Note: It is particularly important that the elements be as clean as possible before starting this treatment. Otherwise, the chlorine demand on the water may exceed the quantity of free chlorine in solution and result in incomplete oxidation of the organic film on the elements.

Treatment 4

1. Fill the unit with clean tap water. Do not precoat.
2. Be sure the return valve to the culture system is closed.
3. Add 8 oz of calgon and 4 oz of laundry detergent (select a brand formulated for use in cold water) per 30 gal of water in the filter.
4. Recycle for 1 hr.
5. Drain the filter and backwash (or hose) the elements three times in succession, letting the unit drain completely after each cleaning.
6. Fill the filter with clean tap water and recycle for 30 min, then drain and backwash (or hose) once more.
7. Fill the filter with clean fresh or sea water, precoat, and resume normal filtration. The vacuum or pressure gauge should indicate zero after precoating No bare spots should be visible on the elements.

Note: Extra backwashing is necessary to remove all traces of residual detergent from the elements.

2.8 EVALUATION OF METHODS

The recommendations made in this section are based on the operating principles given in the preceding pages and are broad enough in scope to apply to most culture situations.

Sand-pressure filters cannot normally be relied on as the only source of biological filtration in a culture system. Both types of rapid sand filters will sustain a population of bacteria that contributes to biological oxidation as long as the units are in continuous operation and not dried out part of the time. Sand-vacuum filters are superior from a biological filtration standpoint, but are only practical in systems larger than 10,000 gal.

Sand-pressure units are excellent mechanical filters as long as they are only used to supplement the biological filter bed. They take up less space than sand-vacuum filters, are cheaper to install, and require less maintenance. Both types of filters have the disadvantage of working parts being buried underneath the gravel.

Both types of diatomaceous earth (DE) filters require more maintenance than rapid sand filters and are impractical on systems of less than 1000 gal. The sleeves must be periodically removed and laundered. The cloth occasionally tears and requires patching or replacing. In nearly every case a DE vacuum filter is superior to a pressure unit because the elements can be easily inspected. Also, the working parts in a vacuum filter are more accessible. Routine checks of pressure filters, however, require shutting down the unit, draining it, and opening the pressure tank.

Diatomaceous earth filters are more expensive to operate on a long-term basis than rapid sand filters. The filter bed in gravel filters is a permanent installation. A DE filter cake, on the other hand, is discarded at the end of a cycle run and must be replaced. The cost of DE can represent a considerable yearly expense in the maintenance of large units in continuous operation.

On the positive side, DE filters provide better clarity than gravel filters by removing greater amounts of organic colloids, particulate matter, and suspended microorganisms per volume of filtered water. DE filters are recommended for systems where these factors must be held to minimal levels.

The operating cost of a DE filter can be reduced by connecting it in series after a rapid sand filter. This arrangement minimizes the particulate load on the elements and prolongs the cycle runs. The same arrangement is especially useful in filtering natural water. Diatomaceous earth filters cannot be operated efficiently if the influent water is highly turbid. The filters become clogged too easily and require frequent backwashing. The same factor limits their use as adjuncts in cleaning biological filters.

CHAPTER 3

Chemical Filtration

3.1 DEFINITION AND FUNCTION

Biological filtration removes a portion of the organics in culture water by mineralization. Mechanical filtration with rapid sand and diatomaceous earth filters reduces the level of particulate material and colloidal organics. But neither process is adequate to keep the level of dissolved organics within safe limits. This can only be accomplished with chemical filtration.

Chemical filtration is the removal of substances (primarily dissolved organics, but also nitrogen and phosphorus compounds) from solution on a molecular level by adsorption on a porous substrate, or by direct chemical fractionation or oxidation. This is necessary for two basic reasons. First, a reduction in the organic level decreases the number of available substrates for heterotrophic oxidation. This action lowers the oxidative potential of the system and helps keep it within its carrying capacity. Second, many of the organics present in solution have unfavorable effects on aquatic animals (see Sect. 7.3).

Many animals, particularly marine invertebrates, actively take up dissolved organics from water. This activity appears to be selective (Fontaine and Chia, 1968) and may be important in normal growth and physiological balance. However, when vitamins are added to a closed system, the portions not utilized by the animals provide additional energy sources for heterotrophic oxidation. Even beneficial organic additives, therefore, ultimately contribute to the rising organic level and a reduction in the carrying capacity of the system.

From a standpoint of culturing, what accumulates in the water in the way of organics is far more significant than what may be lacking. Vitamins necessary for growth can be individually added to culture water, but no chemical filtration process is as selective in the substances it removes. This shortcoming is further complicated by the complexity and number of organic species found in viable culture water, few of which have ever been measured or described. (The most comprehensive results so far have been reported by Deguchi, 1960.)

There are additional reasons why the dissolved organics should be kept at a low level. Many chloroform-soluble substances interfere with normal gas exchange at the air–water interface. These are the higher molecular weight, less soluble fatty acids and alcohols. They force more soluble species out of surface positions when there is competition for adsorption sites (Garrett, 1967). It is significant that in closed systems lipid residues from food and animal excreta often form insoluble surface films. Under conditions of low turnover these compounds can seriously interfere with gas exchange at the surface.

Dissolved organics prolong the life of air bubbles entering culture water from artificial aeration. In marine systems especially, they cause extensive surface foaming and the chronic presence of fine air bubbles in suspension.

Garrett (1967) found that the life span of bubbles at the surface depended partly on the chemical composition of the water—in other words, whether a surface film was present. During tests, bubbles burst immediately upon reaching the surface in sea water containing little organic material. Miyake and Abe (1948) noted that the life of surface foam in sea water was temperature dependent. At higher temperatures the foam dissipated faster. However, the height of the foam depended upon the quantity of organic material in solution. In natural sea water the height was 1.7 cm and in an artificial mix of inorganic salts, 1.2 cm. The difference was attributed to the presence of dissolved organics in the natural medium.

Some basic configurations regarding chemical filtration can be established despite the lack of comprehensive data. These are discussed in the following two sections. In the final section the various methods are evaluated. An excellent review of chemical filtration methods in waste-water treatment has been given by Clesceri (1968).

3.2 REMOVAL OF DISSOLVED ORGANICS BY ADSORPTION

Activated Carbon

Activated carbon, or charcoal, is a porous substance and its degree of efficiency, or adsorptive capacity, is measured by the total surfaces, within the pores, that are available to chemically attract organic molecules. The extent of these surfaces is considerable. One pound of a high-grade powdered variety may contain several million square feet of surface area. In aquatic animal culture it is usually the granular type that is used. An artist's conception of the internal structure of an activated carbon granule is shown in Fig. 14.

An excellent study on the effectiveness of activated carbon in the removal of organics from waste water was made by Parkhurst et al. (1967). Waste water that had not previously been filtered or chemically treated was passed

Chemical Filtration

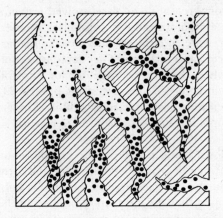

Figure 14. Diagrammatic cross section (magnified) of an activated carbon granule showing adsorbed molecules in the pores.

through large pressure filters containing granular activated carbon. The flow rate through the carbon beds was 7 gsfm.

The results of the tests are given in Table 2. They indicate that activated carbon is an excellent mechanical filtrant, as seen from the reduced values for suspended solids and turbidity (measured in Jackson Turbidity Units).

Table 2. Removal of Contaminants from Waste Water with Activated Carbon

Contaminant	Influent		Effluent
Suspended solids	10.0	mg/liter	<1.0
COD	47.0	mg/liter	9.5
Dissolved COD	31.0	mg/liter	7.0
Total organic carbon (TOC)	13.0	mg/liter	2.5
Nitrate, as N	6.7	mg/liter	3.7
Turbidity	10.3	JTU	1.6
Color	30.0		3.0
Odor	12.0		1.0

The COD (chemical oxygen demand), dissolved COD, and TOC (Total Organic Carbon) were substantially reduced after the water passed through the carbon beds. Color and odor of the filtered water also decreased, giving further indication of the reduction in organic contaminants.[1]

---

[1] Chemical Oxygen Demand is a measure of the oxidizable organic compounds present in water. It is not to be confused with Biological Oxygen Demand (BOD), which is a different process (see American Public Health Association et al., 1965).

The reduced nitrate values of the effluent water were attributed to denitrification by bacteria on the carbon surfaces rather than to adsorption by the carbon. The investigators estimated that a 10–20 per cent removal of the dissolved COD was brought about by biological mineralization. This took place in the first carbon contactor in series. The last carbon bed, in other words, removed the same amount of dissolved organics as did the other three, even though water had already passed through the other beds. Second, and more important, it was found that as much as 55 lb of dissolved COD could be removed per 100 lb of activated carbon.

Additional evidence of the high efficiency of activated carbon in sludge treatment has been given by Weinberger et al. (1966), who state that

. . . After removal of colloidal and suspended solids, the soluble refractory organics may be efficiently removed by contact with activated carbon granules. Such carbon will adsorb up to 20–30 per cent of its own weight in mixed organics from waste water. . . .

Six factors affect the adsorption rates of activated carbon: pH, temperature, concentration of dissolved organics, size of carbon granules, type of carbon used, and contact time between carbon and the water.

A decrease in pH results in the reduced adsorption of negatively charged substances (Morris and Weber, 1964). Presumably, an increase in pH would result in reduced adsorption of positively charged compounds. The pH of culture water is a relatively stable factor, however, and probably does not affect adsorption significantly.

The efficiency of activated carbon increases with increasing temperature (Morris and Weber, 1964). However, temperature, like pH, is a stable factor in well-managed culture water and is therefore not likely to be significant. Nevertheless, it is important to note that adsorption increases with increasing temperature, since this implies that activated carbon may be slightly more efficient in warm-water than in cold-water systems.

The quantity of dissolved organics removed by activated carbon is not a linear function with time. According to Morris and Weber (1964), adsorption involves the ". . . rapid formation of an equilibrium interfacial concentration followed by slow diffusion into the carbon particles. . . ." In other words, the adsorption rate is greatest initially, then tapers off.

The smaller the carbon granules, the greater the surface area. Powdered varieties have the most surface area, as mentioned in the first section. Generally, powdered carbon is impractical because it is hard to handle and difficult to keep out of the culture system itself. Granular types are both chemically efficient and easy to handle.

Many brands and types of activated carbon are available, but no one has yet compared their properties in culture water. Most carbons are manufactured from cellulose-base materials, such as coal, wood, and several kinds of nut shells, notably coconut and pecan. Charcoal is also made from animal bones. This type has the advantage of being more dense than water, but its efficiency seems to be less than that of the others.

The quantity of organics adsorbed is partly a function of contact time. Total adsorption increases with increasing contact time.

Regeneration, or restoration of the adsorptive properties of saturated carbon, can be effected only by steam pressure. Dry heat alone, such as baking in an oven, cannot decompose and drive off a significant portion of the accumulated organics. Parkhurst et al. (1967) gave a description of a regeneration furnace. Considering the expense of regeneration, it is cheaper simply to discard used carbon, except in extremely large installations—that is, those in which more than a million gallons of water is processed each day.

When activated carbon is used in combination with biological filtration, it is best to keep the carbon in a separate container. Scattering it on the surface of a filter bed is impractical, because once the material is saturated it then becomes necessary to separate it from the gravel grains.

In culture systems of 100 gal or less, standard aquarium corner or outside filters packed with granular carbon are efficient. Since water entering these containers had not been prefiltered by the gravel bed, it is best to place tight plugs of glass wool inside the containers on top of the carbon. This reduces colloidal surface coating of the granules.

In airlifted systems with volumes of 100–1000 gal, a carbon contactor can easily be constructed from a length of PVC pipe. Each end of the contactor should be threaded and fitted with removal caps for easier maintenance. The effluent end should have an inset perforated plate, a section of plastic screen, or a plug of glass wool to prevent the carbon from being sucked into the airlift. The end caps are drilled, tapped, and fitted with threaded nipples. The design should incorporate a by-pass arrangement, like the one shown in Fig. 15, to let the water recirculate through the biological filter when the contactor is shut down for recharging. The contactor may be placed in a horizontal position underneath the culture system, as shown in Fig. 16.

The contactor is recharged by first disconnecting it at both ends. It is easier to unscrew the influent and effluent caps if flexible hose is used at the connection points instead of rigid PVC pipe. The used material must be replaced with new carbon which has been thoroughly washed in clean tap water to remove the dust. The threaded male ends can be wrapped with Teflon tape to prevent leaking around the caps and nipples.

For systems of 1000–10,000 gal, suitable contactors can be made from 55-gal drums with removable lids. The insides of the drums should first be painted with two coats of a durable epoxy paint to retard rusting. Two 1-in. holes are then drilled in the sides, one near the top and the other at the bottom. One-inch threaded PVC flanges are sealed against each hole on the

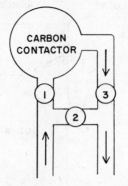

Figure 15. By-pass design for a carbon contactor. For normal filtration valves *1* and *3* are opened and *2* is closed. To recycle through the biological filter, valve *2* is opened and *1* and *3* are closed.

inside, then PVC flexible hose attachments are screwed into the flanges from the outside. A subgravel plate to suspend the carbon above the bottom is unnecessary if a small section of plastic screen is attached to the inside face of the flange. This is adequate to keep the carbon particles from being carried into the airlift. The influent into the contactor should come from a "tee" in the biological filter return line, the same as in Fig. 15. An illustration of a 55-gal steel drum contactor is shown in Fig. 17.

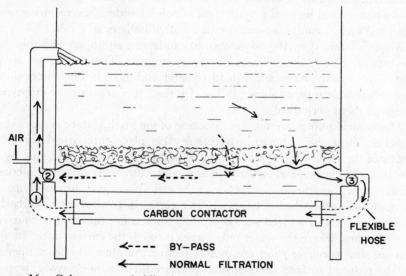

Figure 16. Culture system holding less than 1000 gal showing tank stand, carbon contactor, and by-pass valves. Valve arrangement is the same as in Fig. 15.

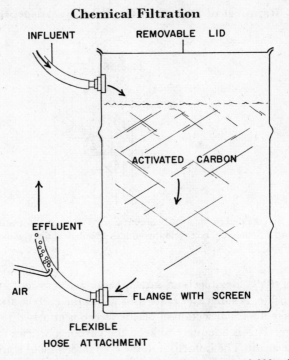

Figure 17. Steel drum contactor for systems 1000–10,000 gal.

Each drum should be filled three-fourths full with washed granular carbon. One drum should be used per 1000 gal of culture water. Several drums servicing one system can be operated either individually or in series.

Systems larger than 10,000 gal require contactors equipped with mechanical pumps. Most sand-pressure filters are suitable. These units are simply filled with granular carbon instead of sand and gravel. The filters should have removable tops, as shown in Fig. 18, since it is necessary to empty and recharge them periodically.

Mechanical pumps are required because of the small surface areas of sand-pressure filters, even though adsorption takes place throughout the entire depth of the carbon bed. A mechanical pump offsets this deficiency by increasing the turnover rate through the contactor. This in effect increases contact time, because in closed systems the same water is continuously passed through the carbon bed. Thus the water comes in contact with the carbon more frequently in a given period of time. Such is not the case, of course, in open systems, where the water only passes through the carbon once. In open systems slow turnover rates are more effective. Another way to compensate for the reduced surface areas in sand-pressure filters is to decrease the mesh size of the carbon. This increases the total surface area of the carbon bed.

The turnover rate should be adjusted so that the entire volume of the system passes through the carbon once in 24 hr.

The basic question of chemical filtration with activated carbon is how often should the filtrant be changed—or to put it another way— how quickly does it become saturated. Estimates range from once a week to never. Both answers contain elements of truth. Earlier in this section it was stated that adsorption rates were not linear, but instead reached high peaks initially and then tapered off. The culturist responsible for only a few small systems may find weekly changes a desirable limit. But if he operates many large systems, the man-hours necessary to service the contactors, plus the expense of the carbon, make such frequent changes impractical.

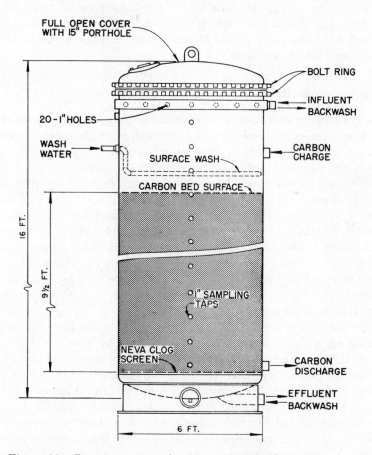

Figure 18.   Pressure contactor for systems 100,000 gal or more.

At the other extreme—never changing the carbon at all—mineralization processes by microbes on the granules ultimately aid in the removal of organics long after the carbon has lost its adsorptive properties. But this could also be achieved by increasing the surface area of the biological filter.

The only sure way of knowing when a carbon bed is saturated is to monitor the organics in the contactor effluent. When they start increasing, it is time to change the carbon.

The minimum quantity of carbon required to maintain dissolved organics in culture water at a specific level has never been determined. Many variables are involved—carrying capacity, the types and rates of prefiltration (see Sect. 4), the nature of the organic contaminants, temperature, pH, and the mesh size of the carbon, to name a few. Thus, the culturist is left very much on his own in deciding how much filtrant to use and how often to change it. In the various contactors described above, 3 months appears to be a workable limit, but it should be emphasized that this is merely an estimate, more wet-thumb than scientific.

Ion Exchange Resins

Ion exchange resins are electrochemically charged resin beads that remove an ionic species from solution by exchanging it with another ionic species. According to Kunin (1963), these resins are manufactured with the following properties: strongly acidic cation, weakly acidic cation, strongly basic anion, and weakly basic anion. The choice of a resin depends upon the properties of the substance to be extracted from solution. Weaker resins are preferable, if they will remove the desired substance, because they are easier to regenerate (eventually a resin becomes exhausted and must be recharged).

Strongly acidic cation exchange resins in the sodium form have proved effective in the removal of ammonia from sewage treatment plant effluent. This has been reported by Nesselson (1954) and also by Culp and Slechta (1966), who used this resin type to achieve 82–99.5 per cent ammonia removal from a plant effluent.

Strongly basic anion exchange resins in chloride form are suitable for nitrate ion removal. This has also been reported by Nesselson (1954). Eliassen et al. (1965) were able to remove 92 per cent of the nitrate and 95 per cent of the phosphate from a sewage treatment plant effluent using this type of resin. The resin could be regenerated with 10 per cent sodium chloride. Regeneration was accomplished with two "bed volumes" (one BV = the volume of the resin bed) of 10 per cent NaCl. Martinez (1962) achieved 99 per cent nitrate and 98 per cent orthophosphate removal from a sewage plant effluent using the same resin type.

It is important to note that all the above tests were conducted in fresh water. It is unlikely that similar results could be attained using sea water with its large concentration and variety of ions. In the case of a sodium cationic exchanger, the sodium ions in solution would quickly occupy the majority of the exchange sites. The same thing would happen with chloride ions in contact with a basic anionic exchange resin. Eliassen et al. (1965) note that

. . . Because the process depends on chloride exchange, a high chloride content in the influent effectively reduces the exchange and, therefore, [nitrate and phosphate] removals. . . .

They conclude that

. . . Effluent quality, therefore, depends more on the initial concentration of chlorides than on the concentration of nitrogen and phosphorus. . . .

The same principle applies to other interfering anions, and Eliassen et al. (1965) go on to say that

. . . Sulfates are exchanged very strongly by the [basic anionic] resin and a large sulfate concentration utilizes the resin capacity before adequate phosphate and nitrate removal can take place. . . .

In sea water, chloride is present in the quantity of 19,000 ppm and sulfate in the quantity of 885 ppm (see Table 8). These same authors noted that a chloride content of only 200 ppm present in their test water caused significant interference with the ion exchange process.

In aquatic animal culture, ion exchange resins have three serious drawbacks. First, they are limited to use in fresh water. Second, they are subject to organic fouling. Third, several of the substances necessary to remove organics from the resin surfaces are either directly toxic or produce unfavorable changes in the water.

Organic colloids in the water coat resin particles and reduce the number of exchange sites. The mechanism appears to be identical with that observed in the coating of mineral carbonate particle (see Sect. 4.3). Organic fouling of ion exchange resins has been reported by Frisch and Kunin (1960) and by Eliassen et al. (1965). Evidence of fouling during phosphate removal has been given by these last authors and is shown in Fig. 19. The contaminating substances could be removed from the resin surfaces with sodium hydroxide, hydrochloric acid, and methanol and by backwashing the resin beds with bentonite, which acts as a mechanical scrubber at the particle surfaces. Bentonite was the most effective mothod, as seen in Fig. 20.

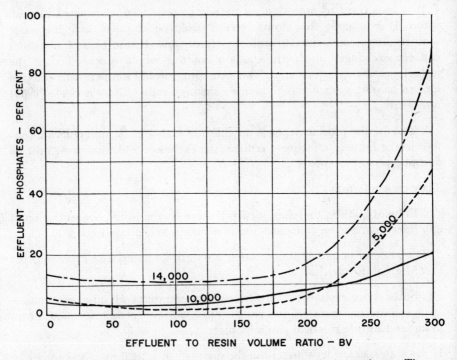

Figure 19. Average removal of phosphates from waste water by ion exchange. The upward shift of the curve indicates decreased efficiency in phosphate removal. The shift to the left indicates reduced adsorptive capacity of the resin due to organic fouling. Numbers on the curves represent continuous flow in bed volumes (BV).

The potential danger of these cleaning techniques is obvious. Excess amounts of NaOH and HCl left in the resin bed after backwashing can rapidly alter the pH, especially in small, poorly buffered systems. Methanol is directly toxic. Only bentonite is safe. However, if bentonite is used, provision must be made to mechanically filter out any leftover material or else the turbidity in the system will increase.

Eliassen et al. (1965) recommended using a diatomaceous earth filter to remove organic colloids before they reached the resin bed, a procedure that would also be necessary in culture application.

There is no conclusive evidence that either nitrate or phosphate is directly toxic to aquatic animals. However, both are considered "nutrients" and can initiate heavy algal blooms in closed systems if allowed to accumulate. Ion exchange resins are potentially useful in nitrate and phosphate removal, although in well-managed culture water the nitrate level seldom exceeds 20 ppm.

Ammonia is extremely toxic to aquatic animals (see Sect. 7.2). In systems where the carrying capacity is stable and few animals are ever added or subtracted, the ammonia level (as total NH_4^+) is seldom greater than 0.1 ppm. However, in commercial culture systems with a continuous turnover of animals the ammonia may exceed this value, especially if the systems are maintained near their maximum carrying capacities. Ion exchange resins may be useful in these situations to remove excess ammonia while the nitrifiers in the filter bed equilibrate.

The work by Eliassen et al. (1965) suggests that ion exchange resins are not very effective in the removal of dissolved COD. The quantity of this factor extracted (and others already mentioned) is summarized in Fig. 21. It is possible—but not likely—that other types of resins might be more suitable than the ones tested. Dissolved organics coat resin particles instead of undergoing direct ionic exchange, and it would appear that after the initial monomolecular layer of organics has formed at the resin surfaces, no further adsorption could be significant.

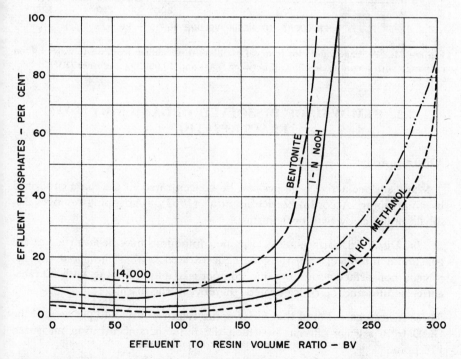

Figure 20. Regeneration of ion exchange resins by various methods. Numbers on the curves represent continuous flow in bed volumes (BV).

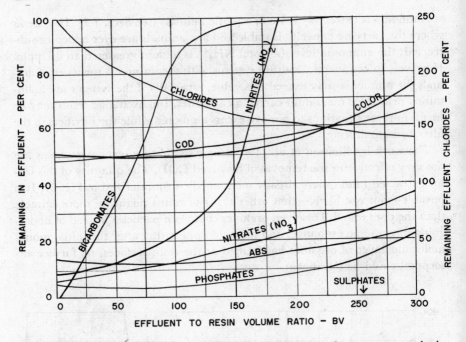

Figure 21. Average values for removal of different substances from waste water by ion exchange with a continuous flow of between 5000 and 10,000 bed volumes (BV).

3.3 REMOVAL OF DISSOLVED ORGANICS BY FOAM FRACTIONATION

Airstripping

Many surface-active organics can be concentrated in the foam produced by airstripping. According to Rubin et al. (1963), airstripping removes dissolved organics by two mechanisms:

. . . (a) Dissolved, surface-active organic compounds may be adsorbed at the gas–liquid interface, and, thus, be concentrated in the foam, and (b) dissolved, nonsurface-active organic compounds may combine with surface-active solutes and be concentrated, thereby, in the foam. . . .

These investigators found that airstripping removed up to 40 per cent of the COD from sewage effluent in which the influent concentration measured 118 ppm.

The efficiency of airstripping depends upon the contact time between the air and the water. Contact time, in turn, depends upon the flow rate of the water into and out of the column, the height of the column, and the volume of injected air. Contact time is increased by decreasing the flow of water through the column and also by lengthening the column. The simplest air-stripping equipment consists essentially of a vertical pipe powered by an air-lift. Since air rises, greater contact time with the water is achieved in long columns.

In airlifted airstrippers, the volume of injected air is the same as would normally be used in an airlift pump of similar dimensions (Table 1). Per cent submergence, as an efficiency factor, is also the same as in airlifts, and the water should emerge from the airstripper in a smooth stream. Contact time is lost if the volume of injected air exceeds the capacity of the lift pipe.

Kuhn (1956) succeeded in removing ammonia from sewage effluent by airstripping. His configurations were: column height = 7 ft, column diameter = 8 in, air volume = 52–55 cfm (cubic feet per minute), and flow rate = 0.10 gpm. The apparatus removed 92.3 per cent of the ammonia from the water. Culp and Slechta (1966) were able to remove 98 per cent of the ammonia from water on a laboratory and pilot plant scale using airstripping. It is significant that in Kuhn's studies the pH of the water was 11. At this high value most of the ammonia is present in the un-ionized (toxic) state (see Sect. 7.2).

There is evidence that airstripping may increase the pH of water. In the sewage effluent tested by Rubin et al. (1963), the average pH prior to airstripping was 7.3. After airstripping, the average value increased to 7.8, while that of the foam in the collection chamber decreased to 7.1 (Table 3). Rubin and his co-workers attributed the increase in pH of airstripped water to the removal of weakly acidic substances from solution. Airstripping is a promising technique for maintaining a stable pH in culture water.

Table 3.   PH Changes in Waste Water after Airstripping

	Range	Mean
Filtered sewage effluent	7.1–7.4	7.3
Water after airstripping	7.6–8.0	7.8
Foam	6.9–7.3	7.1

Several airstripping devices are available for use in small systems. Three types for use on a small scale have been described by Sander (1967). Two of these designs—direct-current and counter-current—are adaptable to much larger systems. Sander refers to these devices as "protein skimmers," a term meant to describe their ability to remove organic substances from solution.

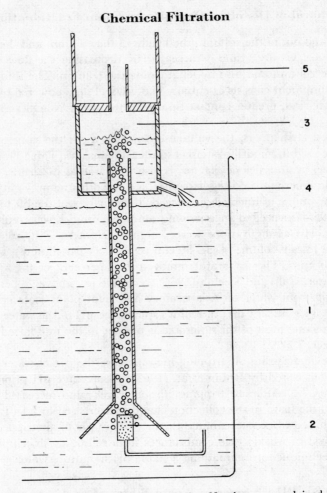

Figure 22. Direct-current airstripping device. Numbers are explained in the text.

A device using the direct-current method is shown in Fig. 22. Air from a compressor is injected through a diffuser (2). As it rises in the column (1) it mixes with the water. Oxygen in the air then oxidizes and coagulates part of the dissolved organic matter. The separation chamber includes a dry upper portion (3).

The coagulated material builds up as a frothy scum on the surface of the water in this chamber. When it has accumulated in sufficient quantity, it is forced through a connecting tube into another, completely dry container (5). This container is periodically removed and cleaned. Excess water is returned to the system from the bottom of the separation chamber through a return line (4).

In the counter-current type, water moves in the opposite direction of the air flow. Part of the air moves downward, instead of all of it moving upward, as it does in the direct-current type. This increases the contact time between air and water and increases the efficiency. Two designs are shown in Figs. 23a and b. In Fig. 23a air moves from the compressor through the diffuser (4) and into the contact tube (2). Water from the culture system enters the contact tube near the top (3). In this design, the top of the tube (2) also serves as the separation chamber. The oxidized material passes into the second (collection) chamber (1), which can then be removed and cleaned. Excess water is removed from below the surface instead of from above, as it is in the direct-current type. Water passes from a connecting tube (6) near the bottom of the contact tube and is airlifted back into the system (5). The water entering the contact tube at 3 passes downward against the air stream and the contact time is increased.

In Fig. 23b the mechanism is essentially the same, except that the contact tube (2) is fitted with a larger outside tube (3) forming a sheath. The advantage is that water in the contact tube cannot be driven back into the system by the injected air, as it can in Fig. 23a at point 3. This makes Fig. 23b slightly more efficient than Fig. 23a.

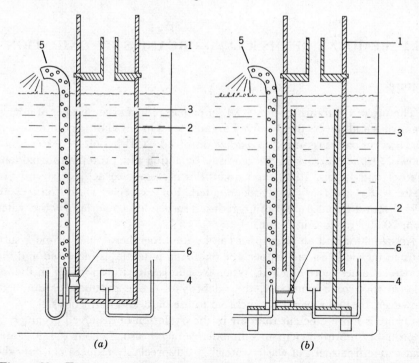

Figure 23.   Two counter-current airstripping devices. Numbers are explained in the text.

Any airstripper is more efficient if the air entering the column is diffused at the injection point. Diffused air produces smaller, more numerous bubbles. As the total surface area of the bubbles increases, so does the electrostatic potential between the bubbles and charged organic compounds. In small devices such as the ones just described, airstones are efficient diffusers. These are shown in Figs. 22 and 23 at the air-injection points. In large columns, such as the one used by Kuhn (1956), the inside of the contact tube can be packed with Rashchig rings. If this is inconvenient, a baffle of perforated PVC pipe can be inserted inside the tube at the air injection point. Either type of diffuser functions well, but Raschig rings are probably more efficient.

Airstripping does not remove particulate organic matter. The device should follow the mechanical filter in series. In most systems this means the filter bed, since the gravel serves both as a biological and a mechanical filtrant. Airstripping is probably more efficient in the removal of dissolved organics than ion exchange resins, but less efficient than activated carbon. Like activated carbon, but unlike ion exchange resins, airstripping can be used in either freshwater or marine culture systems, although it seems to be more efficient in the latter.

3.4 REMOVAL OF DISSOLVED ORGANICS BY OXIDATION

Ozone

The use of triatomic oxygen (O_3), or ozone, reduces the number of microorganisms and lowers the dissolved organic level in circulating culture water. The level of ozone required to oxidize dissolved COD in culture water is not known. Ozone destroys microorganisms by acting as a protoplasmic oxidant (Fetner and Ingols, 1959), and quantities necessary to achieve sterility in waste water are fairly well documented. For example, Dickerman et al. (1954) found that 1.5 ppm of O_3 reduced suspended bacteria in waste water from 70,000/ml to 0 in 5 min.

An ozonator and an airstripper can be used together. When ozone is substituted for air in an airstripper, the oxidizing potential is increased and the device becomes more efficient. When oxygen is substituted for air in the intake of the ozone generator, the quantity of ozone produced is doubled. However, bottled oxygen is expensive to use on a long-term basis.

Ozone is more efficient than air in the oxidation of dissolved organics. It is probably more efficient in ammonia removal also. Lebout (1959) states that ozone treatment of waste water ". . . appreciably reduces not only the proportion of free ammonia in the water, but also the proportion of albuminoid ammonia. . . ." Unfortunately, he gives no figures.

Ozonators for use in small systems (less than 200 gal) are available on the market at a reasonable price, but commercial units for systems of several thousand gallons represent a capital expenditure. The actual generation of ozone is relatively inexpensive, however, and even the large generating equipment can be operated at a nominal cost—$8 per million gallons of treated water.

Ozone reacts with unsaturated organic compounds at the carbon–carbon double bonds to form ozonides. Ozonides are formed by the addition of all three oxygen atoms at a double or triple bond. When these bonds are broken, aldehydes, ketones, and acids are formed. Ozone apparently reacts more readily with carbon–carbon double bonds than with carbon–nitrogen double bonds, but it is not effective in oxidizing saturated compounds (Smith and Cristol, 1966). Functional groups oxidized by ozone are —SH, +S, —NH$_2$, —OH, and —CHO.

Several factors affect the efficiency of ozone. The two most critical are temperature and pH. Ozone becomes more unstable when either of these factors increases. As in other forms of chemical filtration, contact time is important. Ozone is unstable in water and contact time is more critical than the quantity of O$_3$ used. The principles of ozonation are similar to those in airstripping. In most ozonators, untreated water enters a column where it is isolated and constricted into a small space. Ozone is injected into the column and mixed with the water. During treatment, the water–ozone mixture rises in the column just as air rises in an airstripper. Contact time is therefore largely a function of column height, and efficiency increases with longer columns. As a safety precaution, ozone should always be injected into a column where it can dissipate away from the animals. It should never be bubbled freely into the culture system.

Ozone has been used in waste waters to remove manganese from solution (Bean, 1959). Conceivably, prolonged ozonation of marine systems could deplete this element and perhaps others. Partial water changes of 10 per cent every 2 weeks should be given to ozonated systems, especially marine systems, to allow for possible depletion of trace elements.

Ultraviolet Irradiation

The use of ultraviolet lamps reduces the number of microorganisms in circulating culture water in a manner similar to that of ozone. No studies have yet been published comparing the two processes. However, Benoit and Matlin (1966) note that when uv lamps are suspended directly over water, part of the oxidizing efficiency generated by uv irradiation results from ozone production at the air–water interface.

Herald et al. (1962) found that bacteria suspended in sea water at the Steinhart Aquarium were reduced by as much as 99 per cent after passing the water through ultraviolet sterilizers at a rate of 8 gpm. Similar results were obtained by Shelbourne (1964) with marine culture water. Herald et al. (1962) suggested that in some species of marine bacteria the effects of uv irradiation were bacteriostatic rather than bactericidal, since prolonged culture of test plates showed the growth of an unidentified organism after about a week. The same plates had appeared sterile after 48 hr.[2]

Burrows and Combs (1968) reported substantially lowered incidence of disease in cultured salmon after irradiation of the water with uv. The process destroyed organisms smaller than 15 μ, including bacteria, protozoans, and viruses.

Ultraviolet irradiation, like other chemical filtration processes, becomes more efficient with increased contact time. The turbidity in the influent water also determines the sterilizing efficiency of uv. In a test performed by Burrows and Combs (1968), prefiltration of raw river water by rapid sand filters doubled the efficiency of uv equipment that consisted of

. . . an 18-watt ultraviolet radiation unit divided into three series of six 40-watt lamps each, with a capacity of 42 gallons per minute per series.

3.5　EVALUATION OF METHODS

Biological filtration devices are necessary for any culture system. The same gravel bed also mechanically filters the water and this is often sufficient to keep turbidity low. Diatomaceous earth (DE) filters can be used in series after gravel beds in large systems to remove particles smaller than 30 μ. If we consider that biological filtration represents the *primary* (first) filter and DE represents the *secondary* filter, then additional filtration is carried out by *tertiary*, or third stage, equipment. This normally consists of chemical filtration devices.

According to available data, adsorption on activated carbon is the most reliable and efficient means of reducing dissolved organics in culture water. Ion exchange resins are questionable, even in freshwater systems. Adsorption, as a process, is more reliable than fractionation or oxidation because it is not limited to a great extent by the properties of the substances being re-

---

[2] The current unit at Steinhart Aquarium is a product of Steroline Systems Corporation, Santa Fe Springs, California.

moved. As Morris and Weber (1964) pointed out, foam fractionation by air-stripping requires that the substances removed be surface active. Oxidative processes are limited because many compounds are not readily oxidized; for instance, ozone is incapable of oxidizing saturated compounds.

Fractionation or oxidation equipment should follow the mechanical filter in series and precede the carbon bed. The initial removal of surface-active compounds prolongs the life of the activated carbon and reduces the replacement cost. A flow diagram incorporating methods for the removal of organics discussed in the first three chapters is shown in Fig. 24.

A sterilizing device of some type is necessary in large, heavily loaded water systems to minimize the chances of epizootic diseases. At the present time too little is known about the effects of either ozone or uv irradiation to recommend one in preference to the other.

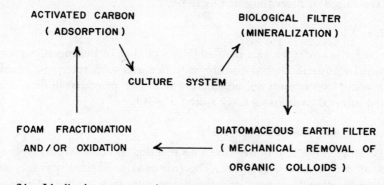

Figure 24. Idealized water processing arrangement incorporating biological, mechanical, and chemical filtration for the removal of organics.

CHAPTER 4

The Carbon Dioxide System

4.1 DEFINITION OF TERMS

The chemical interactions of free carbon dioxide, water, and mineral carbonates constitute the *carbon dioxide system*. The following sections define the terms used in describing its functions.

Buffer

A buffer is any substance that inhibits a change in hydrogen ion concentration. In aquatic animal culture, a buffer acts as a reserve of alkalinity when acidic compounds are added to water. The principal buffers in water are the mineral carbonates $CaCO_3$ and $MgCO_3$.

Alkalinity (= Alkali Reserve)

Alkalinity is the sum of negative ions reacting to neutralize hydrogen ions when an acid is added to water. The two most important negative ions in the carbon dioxide system are carbonate and bicarbonate. Alkalinity is normally expressed as milligrams per liter of equivalent calcium carbonate.

Hardness

Hardness is an expression of the total concentration of calcium and magnesium ions in fresh water and is also expressed as milligrams per liter of calcium carbonate. Hardness is a measure of cations. Calcium and magnesium are the two dominant cations affecting hardness in fresh water; the effects of the others are negligible. Sodium and potassium need not be considered as hardness factors because of their high solubilities. Others, such as iron, copper, and zinc, are usually present only in trace amounts. Fresh waters containing low concentrations of calcium and magnesium are classified as *soft*.

pH

pH is a measure of the hydrogen ion concentration resulting from changes in alkalinity. It is defined as $\log 1/\alpha H^+$. For our purposes, water with a pH of less than 7 at 25 C is acidic; greater than this, basic. A pH of exactly 7

60

is neutral. When one of the dissociation products of a reaction is H+, the solution is acidic and there is a decrease in pH. When a reaction produces OH− ions there is a rise in pH.

Carbonate and Bicarbonate Ions

Carbonate (CO_3^{2-}) and bicarbonate (HCO_3^-) ions are the primary buffers in water. They neutralize any addition of withdrawal of free carbon dioxide and maintain a constant pH by repressing fluctuations in hydrogen ion concentration. The bicarbonate ion is the dominant constituent in the carbon dioxide system within the natural pH ranges of most waters. This is shown in Tables 4 and 5.

Table 4. Percentage Molar Fractions of Carbonic Acid, Bicarbonate, and Carbonate in Cold and Warm Sea Water at Different pH Values (Chlorinity = 19 ‰).

pH	H_2CO_3 (CO_2)	HCO_3^-	CO_3^{2-}
	Temperature 8 C		
7.4	4.9	93.5	1.6
7.5	3.9	94.0	2.1
7.6	3.1	94.1	2.8
7.7	2.5	94.2	3.3
7.8	2.0	93.9	4.1
7.9	1.6	93.2	5.2
8.0	1.2	92.2	6.6
8.1	1.0	91.0	8.0
8.2	0.7	89.3	10.0
8.3	0.6	87.2	12.2
8.4	0.5	84.4	15.1
	Temperature 24 C		
7.4	3.7	93.7	2.6
7.5	2.9	93.9	3.2
7.6	2.3	93.7	4.0
7.7	2.0	93.2	4.8
7.8	1.4	92.5	6.1
7.9	1.1	91.4	7.5
8.0	0.9	89.7	8.4
8.1	0.7	87.9	11.4
8.2	0.5	85.3	14.2
8.3	0.4	82.4	17.2
8.4	0.3	78.9	20.8

TABLE 5. Percentage Molar Fractions of Carbonic Acid, Bicarbonate, and Carbonate in Cold and Warm Fresh Water at Different pH Values

pH	H_2CO_3 (CO_2)	HCO_3^-	CO_3^{2-}
	Temperature 8 C		
5.0	96.9	3.1	0
5.1	96.2	3.8	0
5.2	95.2	4.8	0
5.3	94.0	6.0	0
5.4	92.5	7.5	0
5.5	91.0	9.0	0
5.6	88.5	11.5	0
5.7	86.0	14.0	0
5.8	83.4	16.6	0
5.9	80.0	20.0	0
6.0	75.8	24.2	0
6.1	71.5	28.5	0
6.2	66.2	33.8	0
6.3	61.5	38.5	0
6.4	55.3	44.7	0
6.5	49.7	50.3	0
6.6	43.9	56.1	0
6.7	38.5	61.5	0
6.8	32.8	67.2	0
6.9	28.0	72.0	0
7.0	23.6	76.4	0
7.1	19.8	80.2	0
7.2	16.4	83.6	0
7.3	13.3	86.7	0
7.4	11.0	89.0	0
7.5	8.8	91.2	0
7.6	7.3	92.6	0.1
7.7	5.5	94.3	0.2
7.8	4.7	95.1	0.2
7.9	3.7	96.0	0.3
8.0	3.0	96.7	0.3
8.1	2.5	97.1	0.4
8.2	1.9	97.6	0.5
8.3	1.5	97.9	0.6
8.4	1.2	98.1	0.7
8.5	1.0	98.1	0.9

(continued)

62

Table 5 (*continued*)

pH	H_2CO_3 (CO_2)	HCO_3^-	CO_3^{2-}
	Temperature 8 C		
8.6	0.8	98.1	1.1
8.7	0.6	97.9	1.5
8.8	0.4	97.7	1.9
8.9	0.4	97.2	2.4
9.0	0.3	96.7	3.0
9.1	0.3	96.0	3.7
9.2	0.2	95.3	4.5
9.3	0.2	94.2	5.6
9.4	0.1	93.0	6.9
9.5	0.1	90.9	9.0
9.6	0	89.4	10.5
9.7	0	86.7	13.3
9.8	0	84.0	16.0
9.9	0	80.5	19.5
10.0	0	76.9	23.1
10.1	0	72.5	27.5
10.2	0	68.0	32.0
10.3	0	62.0	38.0
10.4	0	57.3	42.7
	Temperature 24 C		
5.0	95.9	4.1	0
5.1	94.9	5.1	0
5.2	93.7	6.3	0
5.3	92.1	7.9	0
5.4	90.3	9.7	0
5.5	88.2	11.8	0
5.6	85.5	14.5	0
5.7	82.5	17.5	0
5.8	79.1	20.9	0
5.9	74.7	25.3	0
6.0	70.0	30.0	0
6.1	65.0	35.0	0
6.2	59.9	40.1	0
6.3	54.1	45.9	0
6.4	48.1	51.9	0
6.5	42.4	57.6	0
6.6	36.9	63.1	0

(*continued*)

Table 5   *(continued)*

pH	H_2CO_3 (CO_2)	HCO_3^-	CO_3^{2-}

Temperature 24 C

pH	H_2CO_3 (CO_2)	HCO_3^-	CO_3^{2-}
6.7	31.8	68.2	0
6.8	27.4	72.6	0
6.9	23.0	77.0	0
7.0	18.9	81.1	0
7.1	15.8	84.2	0
7.2	13.0	87.0	0
7.3	10.6	89.4	0
7.4	8.4	91.5	0.1
7.5	6.9	92.9	0.2
7.6	5.4	94.4	0.2
7.7	4.5	95.2	0.3
7.8	3.6	96.1	0.3
7.9	2.9	96.7	0.4
8.0	2.3	97.3	0.4
8.1	1.9	97.6	0.5
8.2	1.5	97.8	0.7
8.3	1.2	97.9	0.9
8.4	0.9	98.0	1.1
8.5	0.6	97.9	1.5
8.6	0.5	97.6	1.9
8.7	0.5	97.2	2.3
8.8	0.4	96.7	2.9
8.9	0.4	96.0	3.6
9.0	0.3	95.3	4.4
9.1	0.3	94.1	5.6
9.2	0.2	93.0	6.8
9.3	0.2	91.6	8.2
9.4	0.1	89.6	10.3
9.5	0	87.2	12.8
9.6	0	84.5	15.5
9.7	0	81.2	18.8
9.8	0	77.0	23.0
9.9	0	72.7	27.3
10.0	0	68.5	31.5
10.1	0	63.7	36.3
10.2	0	57.5	42.5
10.3	0	52.5	47.5
10.4	0	46.3	53.7

4.2 DERIVATION OF CARBONATE AND BICARBONATE IONS

Carbonate and bicarbonate ions are derived from three sources: (1) the reaction of free CO_2 with water, (2) the reaction of mineral carbonates with free CO_2 and water, and (3) bacterial reduction processes.

Reaction of Free CO_2 with Water

Free CO_2 is highly soluble in water. It enters culture systems from the atmosphere at the air–water interface and is naturally present in solution as a by-product of metabolism.

Free CO_2 reacts with water to produce carbonic acid (eq. 6). Carbonic acid then dissociates to release hydrogen ions and bicarbonate ions (eq. 7). Bicarbonate ions can further dissociate to give more free hydrogen ions and carbonate ions (eq. 8). This composite reaction is extremely pH sensitive and shifts to the right as pH increases.

$$\overset{(6)}{} \qquad \overset{(7)}{} \qquad \overset{(8)}{}$$
$$2H_2O + 2CO_2 \rightleftharpoons 2H_2CO_3 \rightleftharpoons 2H^+ + 2HCO_3^- \rightleftharpoons 2H^+ + 2CO_3^{2-} \qquad (6,7,8)$$

In hard fresh waters and in sea water, both of which are well buffered, the carbonic acid–bicarbonate equilibrium (eq. 7) is dominant. This is clear from the preponderance of bicarbonate ions within the pH range 7.5–8.3, as shown in Tables 4 and 5.

Reaction of Mineral Carbonates with Free CO_2 and Water

The second, and most important, source of carbonate and bicarbonate is the reaction of mineral carbonates with water and free CO_2. In hard fresh waters and in sea water, much of the carbonate material which potentially affects pH is bound up as calcium and magnesium carbonates. These mineral carbonates act as reserves of potential bicarbonate ions, ready to dissociate and neutralize any increase in hydrogen ions.

As the water in a culture system gradually becomes more acidic as a result of biological oxidation processes, mineral carbonates are brought into solution by reaction with free CO_2 and water. In the case of calcium carbonate, the dissociation products are free calcium and bicarbonate ions (eq. 9).

$$CaCO_3 + CO_2 + H_2O \rightleftharpoons Ca^{2+} + 2HCO_3^- \qquad (9)$$

Bacterial Reduction Processes

Carbonate and bicarbonate ions are formed as the result of bacterial reduction processes—in culture systems, especially during ammonification and deamination by heterotrophs. If the turnover rate is low, there is localized alkalinization at reaction sites on detritus particles and a rise in pH. This can be expected at the surfaces of filter beds with a heavy detritus accumulation and inadequate circulation. If the pH value at the surface of the bed approaches 9, un-ionized ammonia produced by heterotrophic bacteria reacts with calcium ions and produces a precipitate of calcium carbonate plus ionized ammonia. According to Berner (1968), the reaction is

$$NH_3 + Ca^{2+} + HCO_3^- \rightarrow CaCO_3 + NH_4^+ \tag{10}$$

Free CO_2 produced by the combined respiration of microorganisms and animals reacts with water and un-ionized ammonia to produce carbonate and bicarbonate ions if the pH of the water is within normal values (pH < 9). The reactions, again by Berner (1968), are given by eqs. 11 and 12.

$$CO_2 + NH_3 + H_2O \rightarrow NH_4^+ + HCO_3^- \tag{11}$$

$$CO_2 + 2NH_3 + H_2O \rightarrow 2NH_4^+ + CO_3^{2-} \tag{12}$$

4.3 FACTORS AFFECTING THE SOLUBILITY OF MINERAL CARBONATE

The solubility of calcium carbonate depends upon its degree of saturation in water. This, in turn, is dependent on other factors—pH, the presence of magnesium in solution or bound with calcium in the gravel, free CO_2, polar organic substances in solution which coat carbonate particles, and organic phosphates. Temperature is not a significant factor within the narrow range of culture waters. Grain size of the carbonate particles is not significant because of the rapid rate at which particles are coated by dissolved organics.

pH

Calcium carbonate is more soluble at pH values below 5, since it reacts with free CO_2. Its solubility declines as the pH increases. At stable pH values, calcium carbonate is in equilibrium with the rest of the carbon dioxide system.

Presence of Magnesium

Calcium carbonate exists in nature in three forms: calcite, aragonite, and vaterite (Simkiss, 1964a, 1964b; Chave and Suess, 1967). Calcite is the most stable form and also the least soluble.

Using artificial sea water of different compositions, Simkiss (1964a) found that aragonite was precipitated from a solution containing magnesium. When magnesium was eliminated from the medium, the precipitates were calcite and vaterite in the presence of the other cations sodium, calcium, and potassium. The presence of magnesium in sea water apparently inhibits the formation of less soluble calcite crystals (Simkiss, 1964a, 1964b; Berner, 1966) and probably accounts for aragonite being the major precipitate rather than calcite.

Calcites containing traces of magnesium are more soluble than pure calcite (Schmalz and Chave, 1963; Jansen and Kitano, 1963). Chave and Suess (1967) found that magnesium calcites were more than four times as soluble as pure calcite, with a definite trend towards increased solubility as the percentage of magnesium increased. Using natural carbonate sediments of marine origin and treatments with known amounts of $CaCl_2$ and $MgCl_2$, Berner (1966) determined the gain and loss of Mg^{2+} and Ca^{2+}. He noted a mole-for-mole replacement of cations in solution with those at the surfaces of the sediment particles. In simplified form, the reaction was

$$Mg^{2+} \text{ (soln.)} + Ca \text{ (surface)} \rightarrow Ca^{2+} \text{ (soln.)} + Mg \text{ (surface)} \qquad (13)$$

Berner (1966) also noted that only a portion of the total magnesium present on the particle surfaces was exchangeable, regardless of the Ca—Mg ratio of the solution. There were also indications that magnesium calcite was more reactive than aragonite and that direct ion exchange took place only at the surfaces of calcite containing substantial amounts of magnesium. Also, because of the high concentration of sodium in sea water, as much as 20 per cent of the available exchange sites were occupied by this ion. However, this was not thought to be important enough to interfere with normal calcium–magnesium exchange.

No culture water—not even sea water with its large built-in alkali reserve—can indefinitely withstand the acid-forming processes of closed-system culturing. Eventually the hydrogen ions that are produced exceed the natural quantities of carbonate and bicarbonate, and the pH declines. To prevent this from happening, a source of potential carbonates must always be present in contact with the culture water.

The same gravel that biologically and mechanically filters the water also buffers it. *Calcareous* (containing calcium) *gravel* in the filter bed represents an inexhaustible source of buffer material as it slowly dissolves. In effect, the gravel constitutes the alkali reserve and the buffering capacity of a system is directly dependent upon the solubility of the gravel as a source of carbonate and bicarbonate ions.

Many materials have calcareous substrates and can be used as filtrants. However, the efficiency of any filtrant used as a source of bicarbonate alkalinity is largely determined by its percentage composition of magnesium. Gravels high in magnesium are better buffers. Four materials commonly used as filter gravel are crushed oyster shell, crushed coral rock, limestone, and dolomite. Crushed oyster shell is available at feed stores, where it is sold as poultry grit. Limestone and dolomite are available from quarries and building supply dealers in already graded sizes. Coral gravel is generally difficult to obtain since in the United States most of it comes from south Florida.

Of the four, dolomite is the best buffer. Dolomite is 50 per cent magnesium, as indicated by its formula $[CaMg(CO_3)_2]$. Pure limestone (calcite) is the poorest buffer, since it contains virtually no magnesium. Oyster shell and coral rock are intermediate in buffering ability and either is suitable in culture systems.

In marine systems, 100 per cent of the gravel should be calcareous. This assures a stable pH in the range 7.5–8.3. In freshwater systems at maximum carrying capacity, silica gravel is often inadequate to keep the pH above 7. Silica gravel does not act as a buffer. In freshwater systems in which the water is moderately hard, a mixture of 25 per cent calcareous and 75 per cent silica gravel is usually sufficient to sustain a pH above neutral. In soft waters, a greater percentage of calcarous material may be necessary. However, when mixing gravels, it is still a good practice to use the same grade of each type (2–5 mm). This makes mechanical filtration more efficient.

Free CO_2

The solubility of mineral carbonates is affected by the amount of free carbon dioxide in solution. An increase in free CO_2 results in a decrease in pH and the increased solubility of calcium carbonate. For every molecule of carbon dioxide removed, one carbonate ion is precipitated (Simkiss, 1964b). This is shown by equation 14.

$$2HCO_3^- \rightleftharpoons CO_2 + H_2O + CO_3^{2-} \tag{14}$$

Coating of Carbonate Particles by Dissolved Organics

Under ideal conditions, the solubility of mineral carbonates is influenced by changes in temperature, salinity, and the partial pressure of free carbon dioxide. Chave (1965) showed that coatings formed on carbonate particles by polar organic compounds completely prevented them from going into solution. This occurred in spite of the other factors just mentioned.

When sea water was acidified to pH 7.25 by addition of HCl, no activity of naturally occurring carbonates could be detected, although reagent calcite dissolved readily. In later work (Chave and Suess, 1967), it was determined that natural carbonates were completely dissolved only at pH 3. Laboratory tests indicated that organic surface films developed rapidly on carbonate particles.

Organic Phosphates

Simkiss (1964b) found that the presence of certain organic phosphates in water inhibited the precipitation of mineral carbonates. This was suggested by differences in the ability of natural and artificial sea water to precipitate calcium carbonate. It was indicated that natural sea water contained an inhibitory substance not present in the artificial medium. Simkiss tested three organic phosphates in artifical sea water. Their order of effectiveness in inhibiting calcium carbonate precipitation was: pyrophosphate > adenosine triphosphate > glycerophosphate. The addition of orthophosphate to artificial sea water produced essentially the same effect observed in natural sea water. Passing natural sea water through an anion exchange column removed the inhibitory substance, but incubation with alkaline phosphatase and hydrolysis by boiling did not.

Organic phosphates in culture water increase with time. Not one has yet measured their inhibitory effects on mineral carbonate precipitation in closed systems.

4.4 MAINTAINING pH: FACTORS CAUSING A GRADUAL DECLINE IN pH

Dissolved organics reduce the buffering capacity of the gravel in the filter bed by coating carbonate particles and reducing Ca–Mg exchange sites. They also react directly with mineral carbonates and precipitate them on the detritus. Oxidation processes by bacteria increase the levels of free CO_2 and nitrate, both of which lower the pH. Although mineral carbonates become more soluble with increasing concentrations of free CO_2, this is nullified as an equilibrium effect by the rapid coating of carbonate particles by dissolved organics.

Dissolved Organics

Although it has not been demonstrated experimentally, it is probable that organic substances in solution coat the gravel in a filter bed in the same manner they coat carbonate particles in the sea. The same process probably takes place in both freshwater and marine culture systems. The total effect is a drastic reduction in Ca–Mg exchange sites on the gravel surfaces, with a consequent reduction in the buffering capacity of the filter bed. This reasoning finds support in the common observation that old calcareous gravel from a filter bed temporarily becomes a better buffer material after it has been acid washed.

Dissolved organics in culture systems appear to coat the gravel more rapidly than they can be removed from solution by chemical filtration techniques. It is essential, therefore, that other factors affecting the pH be carefully controlled.

Aside from coating the gravel and reducing exchange sites on the gravel surfaces, the accumulation of dissolved organics also affects the stability of calcium in solution. According to Berner (1968),

. . . Upon fat hydrolysis by bacteria the free fatty acids can react with ammonia to form soluble ammonium soaps which in turn precipitate as insoluble calcium soaps.

This indicates that at least a portion of the available calcium in the alkali reserve is lost to the detritus as time goes on. Such a reaction with a fatty acid has been demonstrated by Berner (1968) (eq. 15). The significance of such reactions in terms of their long-range effects on the carbon dioxide system has not yet been determined. Bacterial oxidation, however, does exceed bacterial reduction processes in old systems, as indicated by the decline in pH. The potentially unfavorable effects of calcium precipitation by organics can be partially overcome by keeping the organics at low levels and by periodically removing excessive surface detritus. These procedures effectively lower the total heterotrophic respiration and CO_2 production by reducing the available energy source.

$$2RCOOH + CaCO_3 \rightleftharpoons Ca(RCOO)_2 + CO_2 + H_2O \qquad (15)$$

Oxidation Processes by Bacteria

All oxidation reactions, including nitrification, are acid forming, resulting in the production of free CO_2. Low turnover and poor surface agitation cause a rise in the level of free CO_2. The combined respiration of microorganisms and animals brings about the situation. In weakly buffered, heavily loaded freshwater systems the result is often a dramatic drop in pH in a short time.

Kern (1960) has shown that the high solubility of free CO_2 in water makes it difficult to drive from solution. When the air to a culture system has been left off for several hours, it will require several more hours after normal turnover has been resumed before the excess CO_2 is driven off. High levels of free CO_2 will not occur if the optimum turnover rate is consistently maintained.

The accumulation of nitrate is accompanied by a reduction in the pH of culture water (Honig, 1934). In well-managed systems, the nitrate level is partly reduced by denitrification. When partial water changes are given at the standard rate of 10 per cent every 2 weeks, the nitrate level seldom exceeds 20 ppm. This amount is not enough to interfere with normal pH balance.

4.5   THE SIGNIFICANCE OF pH

The hydrogen ion concentration in well-buffered culture water is not a limiting factor if changes in pH are gradual. Significant fluctuations in pH (which cause a rise in the level of free CO_2) are a problem only in soft water and this can easily be corrected by adding calcareous gravel to the filter bed.

The acceptable pH range in marine culturing is 7.5–8.3 and freshwater culturing, 7–9. Precise duplication of pH values found in nature is unnecessary in both cases. For example, it is a common practice among freshwater hobbyists to keep tropical fishes in acid water on the assumption that it aids in duplicating the natural environment. Here again it should be remembered that captive animals are under a number of physiological stresses as a result of confinement. Acid water merely gives them another to cope with by increasing the partial pressure of free CO_2 in solution (see Sect. 5.3).

The pH values in the open sea vary little. If the decline in pH in a marine culture system is gradual, its effects on the animals are insignificant. It is generally not a good practice to keep marine animals in water of pH less than 7.5, even though the lower threshold of 7, given by Moore (1958), may cause them no harm in the wild.

There is no conclusive evidence that slow changes in hydrogen ion concentration are directly harmful to aquatic animals. However, hydrogen ion is useful in pinpointing other changes taking place in the water. We have just seen that several factors are responsible for a general decline in pH (Sect. 4). The addition of alkaline substances, like slaked lime or sodium bicarbonate, to acid culture water, as has been the practice in some public aquariums (Atz, 1964a), does not correct the underlying cause of the decline.

For example, if the cause is an increase in the nitrate level, then adding buffering material is of little use. This is also true if the cause is an excess of organics that initiates a bloom of heterotrophic bacteria and the subsequent production of free CO_2. Addition of lime does not reduce the level of organic substrates. In fact, it does nothing beneficial but to temporarily decrease the level of free CO_2.

The factors responsible for lowering the pH in culture water must be controlled directly. The fact that the pH is declining below acceptable limits indicates that changes are occurring in the system that need correcting. The act of adding buffers is treating the effect rather than the cause. In well-managed culture water, acceptable pH ranges are sustained indefinitely by the filter gravel and no additional buffering material is necessary.

PART **II**

Effects of Captive Water on Animals

CHAPTER 5

Respiration

5.1 FACTORS AFFECTING OXYGEN SOLUBILITY

Water is a difficult medium for respiration. At saturation it contains only 3 per cent as much oxygen as air and many times more carbon dioxide. It is a dense, highly viscous substance and the creatures within its realm must work hard to extract the oxygen they need.

The partial pressure of a gas in the atmosphere is directly proportional to the volume it occupies, and the partial pressure of each gas is the same. In water, the relationship between the volume of a gas in solution and its partial pressure depends upon its solubility, or ability to react with water. The solubility of oxygen in a culture system depends upon temperature and salinity, and also upon the rate at which it is brought into contact with the water. This last factor is mostly a function of surface agitation.

Temperature

Temperature and solubility of oxygen are inversely related. As the temperature increases, the oxygen-holding capacity of the water decreases, as shown in Table 6.

Salinity

Salinity and solubility of oxygen are also inversely related. An increase in salinity brings about a decrease in dissolved oxygen. This is also shown in Table 6.

Surface Agitation

Oxygen enters water by direct diffusion at the air–water interface. Since there is normally a greater amount of oxygen in the air, gaseous oxygen diffuses into the water where the concentration is less. This process is facilitated by agitating the surface.

Table 6. Dissolved Oxygen (ppm) in Fresh, Brackish, and Sea Water at Different Temperatures

Temp., C	Chlorinity, ‰										
	0	2	4	6	8	10	12	14	16	18	20
1	14.24	13.87	13.54	13.22	12.91	12.59	12.29	11.99	11.70	11.42	11.15
2	13.84	13.50	13.18	12.88	12.56	12.26	11.98	11.69	11.40	11.13	10.86
3	13.45	13.14	12.84	12.55	12.25	11.96	11.68	11.39	11.12	10.85	10.59
4	13.09	12.79	12.51	12.22	11.93	11.65	11.38	11.10	10.83	10.59	10.34
5	12.75	12.45	12.18	11.91	11.63	11.36	11.09	10.83	10.57	10.33	10.10
6	12.44	12.15	11.86	11.60	11.33	11.07	10.82	10.56	10.32	10.09	9.86
7	12.13	11.85	11.58	11.32	11.06	10.82	10.56	10.32	10.07	9.84	9.63
8	11.85	11.56	11.29	11.05	10.80	10.56	10.32	10.07	9.84	9.61	9.40
9	11.56	11.29	11.02	10.77	10.54	10.30	10.07	9.84	9.61	9.40	9.20
10	11.29	11.03	10.77	10.53	10.30	10.07	9.84	9.61	9.40	9.20	9.00
11	11.05	10.77	10.53	10.29	10.07	9.84	9.63	9.41	9.20	9.00	8.80
12	10.80	10.53	10.29	10.06	9.84	9.63	9.41	9.21	9.00	8.80	8.61

13	10.56	10.30	10.07	9.84	9.63	9.41	9.21	9.01	8.81	8.61	8.42
14	10.33	10.07	9.86	9.63	9.41	9.21	9.01	8.81	8.62	8.44	8.25
15	10.10	9.86	9.64	9.43	9.23	9.03	8.83	8.64	8.44	8.27	8.09
16	9.89	9.66	9.44	9.24	9.03	8.84	8.64	8.47	8.28	8.11	7.94
17	9.67	9.46	9.26	9.05	8.85	8.65	8.47	8.30	8.11	7.94	7.78
18	9.47	9.27	9.07	8.87	8.67	8.48	8.31	8.14	7.97	7.79	7.64
19	9.28	9.08	8.88	8.68	8.50	8.31	8.15	7.98	7.82	7.65	7.49
20	9.11	8.90	8.70	8.51	8.32	8.15	7.99	7.84	7.66	7.51	7.36
21	8.93	8.72	8.54	8.35	8.17	7.99	7.84	7.69	7.52	7.38	7.23
22	8.75	8.55	8.38	8.19	8.02	7.85	7.69	7.54	7.39	7.25	7.11
23	8.60	8.40	8.22	8.04	7.87	7.71	7.55	7.41	7.26	7.12	6.99
24	8.44	8.25	8.07	7.89	7.72	7.56	7.42	7.28	7.13	6.99	6.86
25	8.27	8.09	7.92	7.75	7.58	7.44	7.29	7.15	7.01	6.88	6.75
26	8.12	7.94	7.78	7.62	7.45	7.31	7.16	7.03	6.89	6.76	6.63
27	7.98	7.79	7.64	7.49	7.32	7.18	7.03	6.91	6.78	6.65	6.52
28	7.84	7.65	7.51	7.36	7.19	7.06	6.92	6.79	6.66	6.53	6.40
29	7.69	7.52	7.38	7.23	7.08	6.95	6.82	6.68	6.55	6.42	6.29
30	7.56	7.39	7.25	7.12	6.96	6.83	6.70	6.58	6.45	6.32	6.19

When the subgravel filter and airlift are used, deoxygenated water which has passed through the water column and filter bed is recycled back to the surface and agitated. More water is brought in contact with the atmosphere when the turnover rate is high. A high turnover is critical in keeping the dissolved oxygen level near saturation.

Surface agitation also drives free carbon dioxide from solution by bringing a greater portion of the respired water to the air–water interface in a given period of time. This enables the system to maintain a level of free CO_2 in close approximation with the quantity in the atmosphere by reducing its partial pressure in solution. A constant turnover rate of 1 gsfm in systems larger than 200 gal keeps the oxygen level at saturation at all temperatures and eliminates oxygen depletion as a factor when other conditions in a culture system become questionable.

Adequate turnover is especially important in warm-water systems because the animals under culture have higher oxygen requirements. The situation in warm water is further complicated by the relatively low solubility of oxygen at higher temperatures. It should also be kept in mind that the filter bed exerts a considerable BOD and competes with the animals for oxygen.

Temperature and salinity, along with other factors, influence the respiration and metabolism of the animals under culture. The influence they exert is complicated by highly diverse physiological requirements and tolerance levels among species, as well as by physical differences among animals of the same species. Because of the complexity of the aquatic biosphere, the nature and degree of these differences are largely unknown. For this reason, fishes will be used as primary examples simply because they have been more intensely studied. However, since invertebrates are subject to the same stresses, the same principles also apply to them.

5.2 TEMPERATURE

Temperature is the most critical factor affecting respiration. Since temperature largely determines not only the amount of oxygen in solution, but also the amount available to the animal, it exerts a profound influence on metabolism. In cold-blooded animals temperature, respiration, and metabolic rate are inseparably linked. Oxygen consumption is one of the indices commonly used to measure metabolism. Invertebrates and fishes are cold-blooded animals and it is important to remember that their metabolic rates are geared to the temperatures at which they are acclimated. Temperature fluctuations make respiration even more difficult in an already hazardous environment.

Effects on Respiration

The respiration rates of most aquatic animals increase with increasing temperature. This has been corroborated by numerous workers. For example, Kanungo and Prosser (1959) noted that active oxygen consumption in goldfish acclimated to 30 C was 359 per cent higher than in fish acclimated at 10 C. Morris (1962) found that warm acclimation doubled the respiration rate in the cichlid, *Aequidens portalegrensis.*

Wells (1935), Fry and Hart (1948), and Kanungo and Prosser (1959) showed that at intermediate temperatures, cold-acclimated fishes have higher respiration rates than do warm-acclimated ones. Kanungo and Prosser found that when goldfish acclimated to 10 C were place in water which was at 20 C, their standard oxygen consumptions were 26 per cent and their active consumptions 10 per cent higher than fish maintained at 20 C and then acclimated to 30 C.

There is evidence that effects of high temperature coupled with low oxygen may exert a greater effect than either factor alone (Wiens and Armitage, 1961). This is in keeping with the common observation that warm-acclimated animals usually succumb more quickly when the air is shut off in their culture system than do cold-acclimated animals in cold water.

Respiration in the aquatic environment depends upon the partial pressure of oxygen in solution and on the amount of water the animal is able to move past its gills. In the fish, ventilation of the gills must change along with changes in dissolved oxygen if the animal is to maintain normal metabolism. At 0 C, gill ventilation is one-half what it is at 37.5 C, because at the higher temperature the solubility of oxygen is nearly one-half (Rahn, 1966). Also, as the breathing rate increases with rising temperature, there is an increase in metabolic rate, creating a higher oxygen demand on the tissues.

Many species can tolerate only narrow temperature changes. Morris (1965) stated that different mechanisms may be present in different species of bony fishes. In one type, attempts to acclimate a warm-water species to a low temperature result in such a drastic drop in metabolism that the animal is unable to adjust; in the other type, acclimation is seldom fatal.

Thermal Acclimation

Several workers, notably Wells (1935), Schlieper (1950), and Saunders (1962), have demonstrated that thermal acclimation in fishes can be more accurately measured in terms of days than hours. Saunders, for example, showed that complete acclimation of even a hardy species, such as the carp, required 48 hr after transfer from 32 C water to water at 36 C. He considered temperature increases in increments of 1 C per 24 hr to be a proper acclimation rate. Tyler (1966) also used this rate in successfully acclimating two species of minnows to higher temperatures. When acclimating them to lower temperatures, the rate used was $\frac{1}{2}$ C per day.

The rate of 1 C per 24 hr is suitable for acclimating aquatic animals to either higher or lower temperatures in culture systems.

Acclimation of newly captured specimens should always begin at the prevailing temperature of the natural habitat. This prevents stressing them with temperatures too near their lethal limits. In both cold and warm waters, temperatures even slightly higher may cause recently captured animals to exceed their maximum metabolic levels. In temperate climates ambient temperatures are seasonal and what constitutes a suitable ambient temperature at one time of the year may be lethal 6 months later. For instance, a largemouth bass captured in winter seldom survives direct immersion in a warm-water system. Direct immersion in warm water is less harmful to a bass captured during the summer when temperatures in its natural habitat are higher.

When fishes are suddenly placed in water of a different temperature, they undergo thermal shock and exhibit characateristic behavioral patterns. These patterns are more or less the same in all fishes. When the new temperature is higher, the fishes may show increased activity, loss of equilibrium (including aimless darting about, surfacing, floating in unnatural positions, increased fin movement, and remaining in temporary, but stationary, positions with the tail elevated), and a general increase in breathing, as measured by movements of the opercula (Hoff and Westman, 1966). When a fish is subjected to sudden colder temperatures, loss of equilibrium, increased breathing rate, and violent convulsions and spasms are seen (Hoff and Westman, 1966).

In general, raising the temperature in a cold-water system containing cold-acclimated animals stresses the animals considerably. There are two reasons for this. First, increasing the temperature increases the animals' metabolism. Second, as the temperature increases, the solubility of oxygen decreases. This process exceeds the rate at which the animals can thermally adjust. The dissolved oxygen may adjust within hours, while complete thermal acclimation of the animals often requires several days.

The effects of lowering the temperature in a warm-water system containing warm-acclimated animals are not as severe. As the temperature decreases, the metabolism of the animals also decreases, resulting in reduced oxygen demand and consumption. Also, the solubility of oxygen increases with decreasing temperature, making more of it available.

Thermal shock is a major cause of death in newly acquired animals. Tyler (1966) has shown that keeping new fishes in their plastic shipping bags and floating the bags in the culture system, even for short periods of time, increases their resistance to thermal stress. But if there are significant differences between the water temperatures in the bags and in the culture system, the practice is obviously of limited value. The temperatures in small bags equilibrate with the culture system in a matter of minutes, while the metabolic rates of the animals remain unchanged.

In cases where the temperatures are nearly the same and the general physiological state of the fishes is good, the floating technique reduces losses. However, if the animals have been subjected to depleted oxygen levels and elevated carbon dioxide and ammonia during transit, more harm may be caused by keeping them in the bags and increasing their exposure to the unfavorable conditions.

The safest method of acclimating new animals is to adjust the temperature in the culture system to the acclimation temperature of the specimens prior to collecting or receiving them. If the animals are purchased, a reliable dealer will provide the acclimation temperature before shipping them.

Once in the system, the animals should be left at this temperature for at least a week. If they show no signs of stress after this period of time, the temperature can be adjusted to the desired level in increments of 1 C per 24 hr.

5.3 OTHER FACTORS AFFECTING RESPIRATION

Carbon Dioxide

The presence of free CO_2 may depress the affinity of fish blood for oxygen. This is known as the *Root effect*. Its importance in the oxygen consumption of aquatic animals has come under attack in recent years, but its effects may be significant in crowded systems with low oxygen tensions and weak buffering capacities.

Basu (1959), working with the brown bullhead, carp, and goldfish, found that oxygen consumption declined with an increase in carbon dioxide. The rate was a linear function with the logarithm of oxygen uptake and increasing carbon dioxide. The decline was greater under conditions of low oxygen tension. Dahlberg et al. (1968) suggested that the significance of the Root effect may have been overstated in previously published work. This was based on their experiments with largemouth bass and coho salmon. In both species abnormally high concentrations of carbon dioxide failed to affect swimming performance to any great extent. Lenfant and Johansen (1966) could not measure any Root effect in tests with the Pacific dogfish shark. However, Black et al. (1954) had previously noted that when 15 species of freshwater fishes were individually sealed in jars, analysis indicated that dissolved oxygen in the respired water was higher when the carbon dioxide tension was increased. This was interpreted as showing reduced ability to consume oxygen at high levels of carbon dioxide. Saunders (1962) found that the depressing effects of carbon dioxide on resting fishes were temporary and that the animals recovered in 3–5 hr. In swimming fishes, the results were permanently increased breathing rates and a fall in effective oxygen consumption in carp, suckers, and bullheads. Like Basu (1959), Saunders found that the drop in oxygen consumption with increasing carbon dioxide was a linear function.

Fishes can be gradually acclimated to increasing quantities of carbon dioxide, a process that increases their oxygen uptake under conditions of abnormally high free CO_2. In nature, animals have sufficient time to adjust to rising carbon dioxide levels except in regions of sudden, heavy, and widespread pollution. In captivity they do not. When the air is shut off in a closed system, there is a sudden increase in the carbon dioxide tension, and the animals ultimately respire beyond the capacity of their environment.

Both in nature and in captivity, high levels of carbon dioxide produced by biological activity are normally found in conjunction with low oxygen levels. The reverse also holds true and a high dissolved oxygen level usually indicates that free carbon dioxide is at a minimum. In culture systems, surface agitation (which drives part of the excess CO_2 back into the atmosphere) and the buffering activity of the calcareous gravel combine to keep it that way.

The Root effect is a questionable influence on the respiration of aquatic animals. It is automatically eliminated as a potential stress factor in culture systems maintained with adequate turnover rates and buffering capacities.

Salinity

Oxygen consumption and respiration rate increase with increasing salinity (Potts and Parry, 1963). The specific gravity in marine systems should never exceed 1.027; in freshwater systems the reading should never be greater than 1.000.

The following factors are not ordinarily thought to be significant in the management of culture water. They are nevertheless considered below for those rare situations in which they could be important.

Size of the Animals

In general, the metabolic rates in aquatic animals are lower in large specimens than in small ones of the same species (Prosser et al., 1957; Wiens and Armitage, 1961).

Age of the Animals

Roeder and Roeder (1964) found a general decrease in respiration rate with age in the freshwater platy and swordtail. Meuwis and Heuts (1957) found virtually no distinction between age and size in carp as far as respiration was concerned. A stunted carp that had been purposely starved for a year was found to have a respiratory quotient that fell within its age group. When this fish was subjected to thermal stress, however, its reactions were similar to the younger specimens in its weight group. There appears to be no general rule with respect to age and respiration rate in fishes.

Photoperiod

Hoar and Robertson (1959) found that goldfish kept under conditions of controlled light (16 hr a day) for 6 weeks were more resistant to sudden increases in temperature than those maintained under light 8 hr a day. The group subjected to the shorter photoperiod was more resistant to sudden decreases in temperature. The magnitude of the effect varied with the season of the year. Evans et al. (1962) found that at 16 C, groups of rainbow trout had a higher metabolic rate when kept under 16 hr of light a day than similar specimens maintained under 8 hr of light. These workers suggested that light acts through the nervous system and affects metabolism in a manner similar to temperature.

6

Salts and Elements

6.1 SALINITY, CHLORINITY, AND SPECIFIC GRAVITY

Sea water differs from fresh water by having a greater quantity of salts. Common measurements of the salt concentration in sea water are salinity, chlorinity, and specific gravity. The remainder of this section is devoted to defining the above terms, stating their interrelationships, giving the factors affecting each, and presenting the normal values.

Definitions

Salinity is a measure of the total salts in a given weight of sea water. It is traditionally defined as the total amount of solid material in grams contained in 1 kg of sea water when all the carbonate has been converted to oxide, the bromine and iodine replaced by chlorine, and all organic matter completely oxidized. Salinity, by definition, is expressed as grams per kilogram or parts per thousand (‰). A salinity of 34 is written 34‰. Sometimes salinity is written 34 S‰ to keep from confusing it with chlorinity.[1]

Chlorinity is a measure of the total halides in a given weight of sea water and is defined as the mass in grams of pure silver necessary to precipitate the halogens in 328.523 g of sea water. Chlorinity is also expressed in parts per thousand (‰). A chlorinity value may be written either as 19‰, or 19 Cl‰.

Specific gravity is the ratio between the weight of a given volume of sea water and the weight of an equal volume of distilled water at 4 C and 1 atm of pressure. When the temperature of a seawater sample is expressed along with specific gravity, the symbol σ_t is used, where σ represents specific gravity and t stands for the temperature in degrees Centigrade. A specific gravity of 1.024 at 20 C would be written $\sigma_{20} = 24.02$.

Relating Factors

Specific gravity is a measure of the density of a volume of sea water, whereas salinity and chlorinity are weight relationships. The specific gravity

---

[1] Salinity is sometimes expressed as "per cent salt," in which case the decimal point is moved one place to the left. A salinity of 34‰ would be equivalent to 3.4 per cent.

of sea water can be calculated if the temperature, salinity, and pressure are known. Salinity and chlorinity are independent of temperature. Approximate determinations of salinity and specific gravity can be made with salinometers and hydrometers, respectively. Precise determinations of salinity are calculated values based on the chlorinity, as seen in equation 16.

$$\text{salinity} = 0.03 + 1.805 \times \text{chlorinity} \tag{16}$$

Salinity, specific gravity, and temperature relationships are summarized in Table 7.

Normal Values

The normal chlorinity of sea water is considered to be 19‰, which equals a salinity of 34.325‰. However, 35.00 S‰ is more convenient and has been adopted as a standard by oceanographers. The normal specific gravity of sea water is considered to be 1.024 at a salinity of 34‰ and a temperature of 20 C.

6.2   FUNCTIONS AND UPTAKE OF ELEMENTS

The ionic composition of culture water plays a vital role in the metabolic processes of aquatic animals. Bowen (1966) lists the main cellular functions of elements as electrochemical, catalytic, and structural. Elements function electrochemically when they serve as metabolic energy sources. It is likely that all essential elements function as enzyme activators and help regulate the rates of biochemical reactions. In this respect, they demonstrate catalytic functions. Many elements are necessary in the synthesis of such substances as proteins and amino acids. Here the function is structural and the element is a necessary constituent in the final product.

Most, if not all, of the known elements are found in natural waters. Many have no measurable effects and probably are not essential. Arnon and Stout (1939) list three factors that determine whether a given element X is essential:

1. The organism cannot grow or complete its life cycle if X is not available.
2. X cannot be completely replaced by another element.
3. X directly influences the metabolic functions of the organism.

Elements enter animals by two mechanism: simple diffusion and active uptake. *Diffusion*, in which an ion moves from a greater concentration in the water into the more dilute cellular fluid needs no explanation. *Active uptake* is the selective extraction of elements from the water against a concentration

Table 7. Corresponding Densities and Salinities of Fresh, Brackish, and Sea Water at 15 C

Density	Salinity	Density	Salinity	Density	Salinity
0.9991	0.0	1.0026	4.5	1.0061	9.0
0.9992	0.0	1.0027	4.6	1.0062	9.2
0.9993	0.1	1.0028	4.7	1.0063	9.3
0.9994	0.3	1.0029	4.8	1.0064	9.4
0.9995	0.4	1.0030	5.0	1.0065	9.6
0.9996	0.5	1.0031	5.1	1.0066	9.7
0.9997	0.7	1.0032	5.2	1.0067	9.8
0.9998	0.8	1.0033	5.4	1.0068	9.9
0.9999	0.9	1.0034	5.5	1.0069	10.1
1.0000	1.1	1.0035	5.6	1.0070	10.2
1.0001	1.2	1.0036	5.8	1.0071	10.3
1.0002	1.3	1.0037	5.9	1.0072	10.5
1.0003	1.4	1.0038	6.0	1.0073	10.6
1.0004	1.6	1.0039	6.2	1.0074	10.7
1.0005	1.7	1.0040	6.3	1.0075	10.8
1.0006	1.8	1.0041	6.4	1.0076	11.0
1.0007	2.0	1.0042	6.6	1.0077	11.1
1.0008	2.1	1.0043	6.7	1.0078	11.2
1.0009	2.2	1.0044	6.8	1.0079	11.4
1.0010	2.4	1.0045	7.0	1.0080	11.5
1.0011	2.5	1.0046	7.1	1.0081	11.6
1.0012	2.6	1.0047	7.2	1.0082	11.8
1.0013	2.8	1.0048	7.3	1.0083	11.9
1.0014	2.9	1.0049	7.5	1.0084	12.0
1.0015	3.0	1.0050	7.6	1.0085	12.2
1.0016	3.2	1.0051	7.7	1.0086	12.3
1.0017	3.3	1.0052	7.9	1.0087	12.4
1.0018	3.4	1.0053	8.0	1.0088	12.6
1.0019	3.5	1.0054	8.1	1.0089	12.7
1.0020	3.7	1.0055	8.2	1.0090	12.8
1.0021	3.8	1.0056	8.4	1.0091	12.9
1.0022	3.9	1.0057	8.5	1.0092	13.1
1.0023	4.1	1.0058	8.6	1.0093	13.2
1.0024	4.2	1.0059	8.8	1.0094	13.3
1.0025	4.3	1.0060	8.9	1.0095	13.5

(*continued*)

Table 7 *(continued)*

Density	Salinity	Density	Salinity	Density	Salinity
1.0096	13.6	1.0131	18.2	1.0166	22.7
1.0097	13.7	1.0132	18.3	1.0167	22.9
1.0098	13.9	1.0133	18.4	1.0168	23.0
1.0099	14.0	1.0134	18.6	1.0169	23.1
1.0100	14.1	1.0135	18.7	1.0170	23.3
1.0101	14.2	1.0136	18.8	1.0171	23.4
1.0102	14.4	1.0137	19.0	1.0172	23.5
1.0103	14.5	1.0138	19.1	1.0173	23.7
1.0104	14.6	1.0139	19.2	1.0174	23.8
1.0105	14.8	1.0140	19.4	1.0175	23.9
1.0106	14.9	1.0141	19.5	1.0176	24.0
1.0107	15.0	1.0142	19.6	1.0177	24.2
1.0108	15.2	1.0143	19.7	1.0178	24.3
1.0109	15.3	1.0144	19.9	1.0179	24.4
1.0110	15.4	1.0145	20.0	1.0180	24.6
1.0111	15.6	1.0146	20.1	1.0181	24.7
1.0112	15.7	1.0147	20.3	1.0182	24.8
1.0113	15.8	1.0148	20.4	1.0183	25.0
1.0114	16.0	1.0149	20.5	1.0184	25.1
1.0115	16.1	1.0150	20.6	1.0185	25.2
1.0116	16.2	1.0151	20.8	1.0186	25.4
1.0117	16.3	1.0152	20.9	1.0187	25.5
1.0118	16.5	1.0153	21.0	1.0188	25.6
1.0119	16.6	1.0154	21.2	1.0189	25.8
1.0120	16.7	1.0155	21.3	1.0190	25.9
1.0121	16.9	1.0156	21.4	1.0191	26.0
1.0122	17.0	1.0157	21.6	1.0192	26.1
1.0123	17.1	1.0158	21.7	1.0193	26.3
1.0124	17.3	1.0159	21.8	1.0194	26.4
1.0125	17.4	1.0160	22.0	1.0195	26.5
1.0126	17.5	1.0161	22.1	1.0196	26.7
1.0127	17.6	1.0162	22.2	1.0197	26.8
1.0128	17.8	1.0163	22.4	1.0198	26.9
1.0129	17.9	1.0164	22.5	1.0199	27.1
1.0130	18.0	1.0165	22.6	1.0200	27.2

(continued)

Table 7   (*continued*)

Density	Salinity	Density	Salinity	Density	Salinity
1.0201	27.3	1.0241	32.5	1.0281	37.7
1.0202	27.4	1.0242	32.7	1.0282	37.9
1.0203	27.6	1.0243	32.8	1.0283	38.0
1.0204	27.7	1.0244	32.9	1.0284	38.1
1.0205	27.8	1.0245	33.0	1.0285	38.2
1.0206	28.0	1.0246	33.2	1.0286	38.4
1.0207	28.1	1.0247	33.3	1.0287	38.5
1.0208	28.2	1.0248	33.4	1.0288	38.6
1.0209	28.4	1.0249	33.6	1.0289	38.8
1.0210	28.5	1.0250	33.7	1.0290	38.9
1.0211	28.6	1.0251	33.8	1.0291	39.0
1.0212	28.8	1.0252	34.0	1.0292	39.2
1.0213	28.9	1.0253	34.1	1.0293	39.3
1.0214	29.0	1.0254	34.2	1.0294	39.4
1.0215	29.1	1.0255	34.4	1.0295	39.6
1.0216	29.3	1.0256	34.5	1.0296	39.7
1.0217	29.4	1.0257	34.6	1.0297	39.8
1.0218	29.5	1.0258	34.7	1.0298	39.9
1.0219	29.7	1.0259	34.9	1.0299	40.1
1.0220	29.8	1.0260	35.0	1.0300	40.2
1.0221	29.9	1.0261	35.1	1.0301	40.3
1.0222	30.0	1.0262	35.3	1.0302	40.4
1.0223	30.2	1.0263	35.4	1.0303	40.6
1.0224	30.3	1.0264	35.5	1.0304	40.7
1.0225	30.4	1.0265	35.6	1.0305	40.8
1.0226	30.6	1.0266	35.8	1.0306	41.0
1.0227	30.7	1.0267	35.9	1.0307	41.1
1.0228	30.8	1.0268	36.0	1.0308	41.2
1.0229	31.0	1.0269	36.2	1.0309	41.4
1.0230	31.1	1.0270	36.3	1.0310	41.5
1.0231	31.2	1.0271	36.4	1.0311	41.6
1.0232	31.4	1.0272	36.6	1.0312	41.8
1.0233	31.5	1.0273	36.7	1.0313	41.9
1.0234	31.6	1.0274	36.8	1.0314	42.0
1.0235	31.8	1.0275	37.0	1.0315	42.1
1.0236	31.9	1.0276	37.1	1.0316	42.3
1.0237	32.0	1.0277	37.2	1.0317	42.4
1.0238	32.1	1.0278	37.3	1.0318	42.5
1.0239	32.3	1.0279	37.5	1.0319	42.7
1.0240	32.4	1.0280	37.6	1.0320	42.8

gradient. It is closely correlated with temperature and a 10 degree rise increases adsorption by 100 per cent (Bowen, 1966). It also depends upon available oxygen; when respiration is inhibited, so is the active uptake of ions from the medium.

6.3   TOXIC EFFECTS OF ELEMENTS

Importance of a Balanced Culture Medium

With few exceptions, pure salt solutions are toxic to aquatic animals. The elements in sea water demonstrate nutritive or life-sustaining properties only in balanced combinations in which ionic antagonism cancels the poisonous effect of any single species. Polyvalent ions are more readily taken up than di- or univalent ons. This is true of both cations and anions. Competition for adsorption sites within a cell occurs among ions of similar properties. According to Bowen (1966),

. . . True competition occurs between pairs of similar ions such as potassium and rubidium or calcium and strontium. In these cases, excess of one ion in the medium depresses the uptake of the other.

Elements in sea water are listed in Table 8.

Toxic Effects of Heavy Metals

The heavy metals (Pb, Hg, Cu, and Zn) are present in water in trace amounts. Nevertheless, all are toxic to aquatic animals in greater, but still very low, concentrations. Many are lethal in quantities of less than 1 ppm.

Zinc is required in normal amounts for several enzymatic functions and is present in many proteins as a structural element. Copper is also found in several enzymes and serves as a respiratory pigment in the blood proteins of many invertebrates.

Zinc and copper are both used to treat protozoan infections in fishes, particularly marine fishes, where the toxic effects of the metal are reduced after a time by precipitation with calcium carbonate. Both zinc and copper cause fishes to produce excessive mucus. The mucus is shed into the water along with the encysted stages of some parasites. It is thought that the swarming stages of parasitic protozoans, such as *Cryptocaryon* (described by Nigrelli and Ruggieri, 1966), are more susceptible to heavy-metal poisoning and that treating a culture system with zinc or copper kills these organisms before they can attach to the host.

Table 8. Elements in Sea Water

Element	Chemical form	Amount, ppm
Ag	$AgCl_2^-$	0.0003
Al		0.01
Ar	Ar	0.6
As	AsO_4H^{2-}	0.003
Au	$AuCl_4^-$	0.000011
B	$B(OH)_3$	4.6
Ba	Ba^{2+}	0.03
Be		0.0000006
Bi		0.000017
Br	Br^-	65
C	CO_3H^-, organic C	28
Ca	Ca^{2+}	400
Cd	Cd^{2+}	0.00011
Ce		0.0004
Cl	Cl^-	19,000
Co	Co^{2+}	0.00027
Cr		0.00005
Cs	Cs^+	0.0005
Cu	Cu^{2+}	0.003
F	F	1.3
Fe	$Fe(OH)_3$	0.01
Ga		0.00003
Ge	$Ge(OH)_4$	0.00007
H	H_2O	108,000
He	He	0.0000069
Hf		<0.000008
Hg	$HgCl_4^{2-}$	0.00003
I	I^-, IO_3^- ?	0.06
In		≪0.02
K	K^+	380
Kr	Kr	0.0025
La		0.000012
Li	Li^+	0.18
Mg	Mg^{2+}	1350
Mn	Mn^{2+}	0.002
Mo	MoO_4^{2-}	0.01
N	Organic N, NO_3^-, NH_4^+	0.5
Na	Na^+	10,500
Nb		0.00001
Ne	Ne	0.00014

(continued)

Table 8 *(continued)*

Element	Chemical form	Amount, ppm
Ni	Ni^{2+}	0.0054
O	H_2O, O_2, SO_4^{2-}	857,000
P	PO_4H^{2-}	0.07
Pa		2×10^{-9}
Pb	Pb^{2+}	0.00003
Ra		6×10^{-11}
Rb	Rb^+	0.12
Rn	Rn	6×10^{-16}
S	SO_4^{2-}	885
Sb		0.00033
Sc		<0.000004
Se		0.00009
Si	$Si(OH)_4$	3
Sn		0.003
Sr	Sr^{9+}	8.1
Ta		<0.0000025
Th		0.00005
Ti		0.001
Tl	Tl^+	<0.00001
U	$UO_2(CO_3)_3^{4-}$	0.003
V	$VO_5H_3^{2-}$	0.002
W	WO_4^{2-}	0.0001
Xe	Xe	0.000052
Y		0.0003
Zn	Zn^{2+}	0.01
Zr		0.000022

All evidence points to the treating of fishes with heavy metals as a dangerous practice of doubtful therapeutic value. In most cases the unfavorable effects on the host far outweigh any benefits gained by killing its parasites. This is true for several reasons. First, no one has yet established the lethal level of a heavy metal, under carefully controlled conditions, in any species of marine fish. Treating a mixed-species system is particularly ludicrous since the long-term effects, and even the lethal thresholds among fishes in general, are known to differ considerably. Moreover, no one has yet shown the effects, under carefully controlled conditions, of a heavy metal on every stage in the life cycle of a single parasitic organism. In short, the treatment may do more harm to the host than to the parasites.

A heavy accumulation of detritus on the surface of a filter bed, in addition to a high concentration of organic material in the water, largely determines the copper levels that can be tolerated by microorganisms and culture animals. There is evidence that organics in activated sludge combine with copper and reduce its toxicity to nitrifying bacteria (Tomlinson et al., 1966). Chelation apparently renders copper less toxic to fishes (Doudoroff and Katz, 1953). Fitzgerald (1963) showed that chelating copper sulfate with citric acid in lake water rendered it 500 times less toxic to bluntnose minnows. Chelation enabled 90 per cent of the copper to remain soluble at pH values of both 6 and 8.5.

Chelated copper is often used to treat marine systems when prolonged soluble levels are desired. However, many bacteria are resistant to copper at prophylactic levels and gradually decompose the organic chelate. This eventually allows the copper ions to combine with carbonates and precipitate. It is important to note that none of the published dosage levels recommended for treating marine systems mentioned organics as a possible interfering factor that might alter the toxic state of the copper. Chelation, as we have just seen, reduces the toxicity of copper to fishes. Presumably it has a similar effect on the parasites.

The mechanisms of heavy-metal poisoning in fishes are well documented. Copper acts by forming insoluble organometallic compounds on the gill surfaces. Ellis (1937) believed that more permanent damage occurred by chemical alteration of proteins within the gills. White and Thomas (1912) showed that copper accumulates in the blood and tissues of marine fishes. Jones (1938, 1939) noted increased breathing rates in sticklebacks immersed in solutions of $Pb(NO_3)_2$, $ZnSO_4$, and $HgCl_2$. This was accompanied by a fall in the rate of oxygen consumption, indicating interference with normal respiration. Increased breathing rate is also a common observation in copper-treated fishes.

Heavy metals precipitate and coagulate mucus on the gill filaments. This inhibits oxygen uptake by reducing available surfaces. There is danger of increased toxicity when the turnover rate is low. Goldfish are far more sensitive to lead nitrate $[Pb(NO_3)_2]$ poisoning in waters of low dissolved oxygen (Westfall, 1945).

Factors Affecting Heavy Metal Poisoning

The toxicity of a heavy metal in culture water depends upon seven factors: pH, dissolved oxygen, temperature, volume of the solution in relation to size of fish, frequency with which the solution is renewed, and synergism with other substances in solution. The seventh—the level of organics—has already been discussed.

The pH of the water may be the most important factor. Fresh waters are more weakly buffered than sea water and this explains the magnified effects of heavy-metal toxicity often seen in treated freshwater systems. As far back as 1915, Thomas observed that heavy metals were far more toxic in distilled and soft waters than in hard, alkaline waters. This has since been confirmed by many workers.

It is often thought that precipitation of copper in alkaline water renders it permanently harmless to aquatic animals. This may not be true, as Doudoroff and Katz (1953) have pointed out. At the time of their writing no evidence existed that copper-sensitive fishes, such as trout, could live indefinitely in water with only the copper precipitate present; there is still no such evidence.

A high dissolved oxygen level nullifies the toxic effects of copper to some extent by making respiration easier. Strong surface agitation in a culture system also prevents the accumulation of free CO_2, which would lower the pH and keep the copper soluble.

Increased temperature increases the toxicity of heavy metal salts to fishes.

Carpenter (1927, 1930) demonstrated that the toxicity of lead salts could be reduced by decreasing the amount of water and increasing the size of the fish. He concluded that lead was detoxified by precipitation with the mucus on the fish.

The frequency with which treated water is renewed is an important factor affecting heavy-metal toxicity. When the water is not changed at all, fishes can partly detoxify it by precipitation of the metal ions with mucus.

Synergistic effects between two heavy metals and between a heavy metal and other substances have been reported by Doudoroff (1952). He found that copper–zinc combinations are sometimes more toxic than either copper or zinc alone. This was later confirmed by Lloyd and Herbert (1962). Herbert and VanDyke (1964) studied the synergistic effects of copper and ammonia. They noted that copper has a high affinity for ammonia, resulting in the formation of cuprammonium ions, of which $Cu(NH_3)_3^{2+}$ predominate. They suggested that cuprammonium is probably equal to copper in toxicity.

6.4 SYNTHETIC SEA WATER

Many brands of synthetic (artificial) sea water are available on the market. Those lacking trace elements, or containing major elements in abnormal ratios, are unsuitable. Natural sea salts which have been dehydrated never again support the variety of life they originally did (Atz, 1964a), and there are no advantages in choosing them over other types. The most important

Table 9. Ionic Composition (ppm) of Instant Ocean® Synthetic Sea Salts at Specific Gravity 1.025 and 15 C

Cl	18,400
Na	10,200
SO_4	2500
Mg	1200
K	370
Ca	370
HCO_3	140
H_3BO_3	25
Br	20
Sr	8
PO_4	1
Mn	1
MoO_4	0.7
S_2O_3	0.4
Li	0.2
Rb	0.1
I	0.07
EDTA	0.05
Al	0.04
Zn	0.02
V	0.02
Co	0.01
Fe	0.01
Cu	0.003

point is this: if the medium includes essential elements in ratios approximating sea water, then minor deviations among brands are secondary to how the medium is handled. Even natural sea water is inadequate if it is not filtered and otherwise cared for properly. One product—Instant Ocean® Synthetic Sea Salts—has proved satisfactory under a variety of conditions. The salt ratios in this mix, as shown in Table 9, approximate natural sea water.[2]

When mixing small volumes of synthetic sea water, it is important to carefully follow the directions on the package. Clean containers made from inert materials should be used for mixing—for instance, spare aquarium tanks, polyethylene cannisters or jerry cans, or plastic garbage cans with tight-fitting lids. Aquarium Systems, Inc. manufactures a special fiberglass mixing container holding 100 gal. The device is equipped with a pump for transferring the finished solution directly to the culture systems. Most domestic

---

[2] Manufactured by Aquarium Systems, Inc., Eastlake, Ohio.

products are hydrated with tap water measured in U.S. gallons, and a device that accurately measures a gallon is a necessary item.

All synthetic mixes containing buffering substances in proportions analogous to natural sea water should be moderately aerated for 48 hr prior to use to stabilize the pH. Finished solutions should be covered to prevent concentration of the salts from evaporation.

Manufactured mixes are normally hydrated in a single step. However, large volumes of synthetic sea water should be mixed in stages, with each batch of salts allowed to mix for 24 hr before adding the next series. The major salts in large volumes are mixed by diluting one salt at a time. Hydration of the salts is easier if the mixing vat is being filled with tap water while the major salts are being added. Minor and trace salts should not be added at this point because they may be precipitated by the initial high concentration of major salts. In fact, it is best not to add the rest of the mix until the specific gravity (with only the major salts) has been roughly adjusted to 1.025 and the water in the vat has been circulated for 24 hr.

The remainder of this section describes techniques for mixing 8000 gal of synthetic sea water at a time. The combined formula given in tables 10–12 is a slightly modified version of the original formula first described by Segedi and Kelley (1964). It has proved suitable for the large-scale culture of a great variety of marine fishes and invertebrates. It should be pointed out, however, that the original formula is not as modern as the one currently under patent and marketed by Aquarium Systems, Inc. (Table 9). The later mix is probably better suited for culturing delicate larvae and for scientific investigations.

When mixing large volumes of synthetic sea water, gallons per se are not directly measured, as when mixing small volumes. The major components are added one at a time while the vat is filling with tap water, as mentioned

**Table 10. Major Components in 8000 gal of Modified Instant Ocean®
Synthetic Sea Salts**

Component[a]	Amount present, lb
Sodium chloride (NaCl)	1840
Magnesium sulfate (MgSO$_4$)	460
Magnesium chloride (MgCl$_2$)	360
Potassium chloride (KCl)	4)
Calcium chloride (CaCl$_2$)	92
Sodium bicarbonate (NaHCO$_3$)	14

[a] All components should be technical grade.

Table 11. Minor Components in 8000 gal of Modified Instant Ocean® Synthetic Sea Salts

Component[a]	Amount present, g
Strontium chloride ($SrCl_2$)	600
Manganese sulfate ($MnSO_4 \cdot H_2O$)	120
Sodium phosphate ($NaH_2PO_4 \cdot 7H_2O$)	120
Lithium chloride ($LiCl$)	30
Sodium molybdate ($Na_2MOO_4 \cdot 2H_2O$)	30
Sodium thiosulfate ($Na_2S_2O_3 \cdot 5H_2O$)	30

[a] All components should be reagent grade.

Table 12. Trace Components in 8000 gal of Modified Instant Ocean® Synthetic Sea Salts

Component[a]	Amount present, g
Potassium iodide (KI)	2.7
Aluminum sulfate ($Al_2[SO_4]_3$)[b]	26.0
Potassium bromide (KBr)	812.7
Cobalt sulfate ($CoSO_4$)	2.7
Rubidium chloride ($RbCl$)	4.5
Cupric sulfate ($CuSO_4 \cdot 5H_2O$)	0.3
Zinc sulfate ($ZnSO_4 \cdot 7H_2O$)	2.9

[a] All components should be reagent grade.
[b] Heat required to dissolve.

above. The quantities of each salt listed in Tables 10–12 are sufficient for 8000 gal of modified Instant Ocean Synthetic Sea Salts at specific gravity 1.025 and temperature 20 C. If a more dilute solution is desired, then more water can be added than the amount required for a specific gravity of 1.025 and the final solution will total slightly more than 8000 gal.

Modified Instant Ocean Synthetic Sea Salts are mixed in three steps, with each step taking a full day to assure thorough mixing and the hydration of all components. The major salts are mixed on the first day, the minor salts on the second, and the trace salts on the third. By the fourth day all components should be evenly distributed throughout the solution, the temperature lowered to the desired level, and the pH stabilized at 8.3. Minor adjustments in specific gravity should then be made, and by the fifth day the solution is ready to use.

Mixing Components

Major components should be technical grade salts packaged in plastic-lined bags. Bags weighing 100 lb are convenient to handle and favorably priced. Water in the mixing vat can be aerated by means of a large airlift, making a mechanical pump for circulating the water unnecessary. A suitable airlift for a mixing vat consists of a straight length of PVC pipe (4-in. diameter) with a hole drilled in it near one end. The size of the hole should be just large enough to make a tight fit for an air line ($\frac{3}{8}$-in. is adequate). No air-dispersion device is necessary at the end of the air line, since the purpose of the lift is simply to keep water circulating in the vat while the salts are being added. A 2-lb lead weight or brick should be attached to the bottom of the airlift to keep it upright.

The mixing vat must have hot and cold running water discharging through a common valve. When mixing the salts, individual components should be dissolved in a small container, such as a 20-gal plastic garbage can, and the concentrated liquid allowed to spill over the edge of the can into the vat below. Some salts may precipitate and others may pile up in corners, undissolved, if the dry components are dumped directly into the vat.

Tap water at 30 C can be added to the can from a garden hose. The end of the hose should be weighted to keep it submerged. If water is added too fast, the salts may not have time to dissolve before they spill into the vat. There should be an in-line thermometer where the hot and cold water merge for regulating the temperature of the tap water. Figure 25 shows an arrangement with the mixing vat, airlift, garden hose, and dissolving can.

Installations using 8000 gal or more of synthetic sea water per month may find it convenient to add sodium chloride to the mix as a concentrated liquid brine, rather than in crystalline form. This can be done by automatic dissolving equipment developed and sold by the International Salt Company, Inc. Sodium chloride is the major solute in sea water. It is expensive and difficult to handle when purchased in manufactured (evaporated), bagged lots. Bulk salt (unbagged) is cheaper and easier to handle, provided that necessary equipment is incorporated into the plant design. Liquid brine can easily be pumped to any part of the culture installation, while moving bagged salt is awkward and time consuming. Also, adding sodium chloride as brine to synthetic sea water is more accurate and efficient than dissolving bagged salt in the mixing vat.

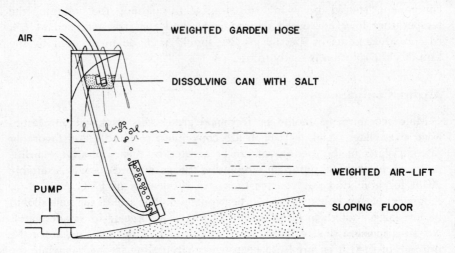

Figure 25. Apparatus for mixing large amounts of synthetic sea water.

A diagrammatic illustration of an automatic dissolver, the Sterling Brinomat,® is shown in Fig. 26.[3] Bulk evaporated salt is stored in the hopper. The hopper is refilled pneumatically through a connecting chute leading outside the plant to a loading ramp for trucks. Tap water entering the Brinomat flows down through the lower portion of the salt bed. As the water moves downward, it forms a brine solution of increasing strength. Just above the bottom it reaches full saturation. Dissolved salt is replaced automatically by dry salt from the hopper. It is important to note that the brine becomes saturated *before* it reaches the bottom of the Brinomat. The lower portion of the bed never dissolves and acts as a mechanical filter which removes insoluble impurities from the effluent brine. As brine is drawn from the Brinomat and transferred to the mixing vat, a float valve opens the tap water line and brine making continues automatically. In marine culture, a Brinomat must be equipped with an accurate flow meter to measure the gallons of brine passing into the mixing vat. Brass meters are preferable, since they are corrosion resistant. The minute amount of copper that may leach from a brass meter is insignificant from a toxicity standpoint. Besides, the tap water used to hydrate the mix will have passed through many feet of copper pipe before it reaches the vat. All-brass, plastic, or stainless steel pumps are recommended for pumping brine or sea water.

---

[3] A product of International Salt Company, Inc., Clarks Summit, Pennsylvania.

Procedure for Mixing Major Components

The quantities of major salts required for 8000 gal of modified Instant Ocean Synthetic Sea Salts are given in Table 10. Sodium chloride, if added from a Brinomat, is normally at saturation and the correct amount is 706 gal. However, the specific gravity of the Brinomat effluent will vary if the hopper is not kept full. The necessary gallons of brine at different specific gravity values are given in Table 13.

Table 13. Gallons of Brine at Different Specific Gravities Necessary to Make 8000 gal of Modified Instant Ocean® Synthetic Sea Salts[a]

Specific gravity of brine	Gallons of brine
1.151	967
1.154	952
1.156	938
1.158	925
1.160	912
1.162	899
1.164	886
1.167	874
1.169	862
1.171	849
1.173	838
1.175	827
1.177	816
1.180	806
1.182	795
1.184	785
1.186	775
1.188	766
1.190	756
1.193	747
1.195	738
1.197	729
1.199	720
1.202	711
1.203 99% saturated	706

[a] Calculations are based on the properties of brine at 15.56 C.

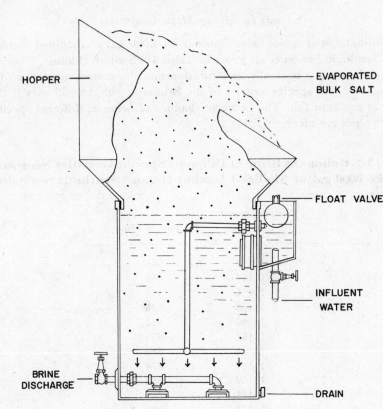

HOPPER

EVAPORATED
BULK SALT

FLOAT VALVE

INFLUENT
WATER

BRINE
DISCHARGE

DRAIN

Figure 26.   The Sterling Brinomat.

1. Be sure the mixing vat is clean. If it is not, scrub the walls and floor with a stiff bristle brush and hose it out thoroughly with tap water. *Do not* use prepared cleansers. Stained areas can be cleaned with a strong solution of sodium bicarbonate and warm water and then hosed off with liberal amounts of tap water.

2. Turn on the airlift.

3. Place the weighted end of the garden hose in the dissolving can. Turn on the water and adjust the temperature to 30 C. Wait until there is about 1 ft of water in the vat before proceeding to the next step.

4. Check the specific gravity of brine from the Brinomat effluent and determine the number of gallons needed from Table 13.

5. Weigh out the correct amounts of the other salts. Fractions of bags should be assembled in clean plastic containers.

6. Turn on the Brinomat pump and add the correct amount of brine.

7. Add the rest of the salts in the order given in Table 10. Use the following procedure:

    a. Fill the dissolving can half-full with a component salt and stir with the stirring stick (a 3-ft length of unpainted dowling) until no salt is left in the can.

    b. Refill the can half-full and repeat *a* above until all salts have been dissolved.

    c. Count the empty bags and containers as a check to be sure that no salts have been forgotten.

8. Continue filling the vat until the water level is just below the 1.025 mark on the side of the vat. (Once the proper water level has been determined for a specific gravity of 1.025, the spot should be marked on the vat wall with epoxy paint.)

Procedure for Mixing the Minor Components

1. One day after the major salts have been mixed, weigh out the proper amount of each minor salt (Table 11). Combine the salts in a large beaker. *Do not* add water.

2. Sprinkle the dry mixture on the surface of the water in the mixing vat.

3. Use the minor salts immediately. If they remain in the beaker for longer than 2 hr, noticeable reactions which alter the chemical states of the components will take place.

Procedure for Mixing the Trace Components

1. Fill a clean 5-gal plastic jug with $\frac{1}{2}$ gal of distilled water.

2. Accurately measure out each salt in Table 12 and place each one in an individual, labeled beaker. Add sufficient water to each beaker to dissolve the salt.

3. Add each solution to the jug. Rinse each beaker with distilled water from a wash bottle and add this water to the jug also.

4. Add distilled water to make approximately 12 liters. Place an air-stone in the jug until the solution is needed. Add the trace elements 1 day after the minor salts have been added.

CHAPTER 7

Toxic Metabolites

7.1 NATURE OF THE EFFECTS

Even at sublethal levels, the toxic metabolites of aquatic animals have at least four adverse effects: (*1*) they increase the susceptibility of the animals to other unfavorable conditions (fluctuating temperature, lack of oxygen, etc.), (*2*) they inhibit normal growth, (*3*) they decrease fecundity, and (*4*) they decrease resistance to disease.

7.2 AMMONIA

Origin of Ammonia in Culture Water

The ammonia in culture water originates from the mineralization of organic substances by heterotrophic bacteria (see Sect. 1.1) and excretion by the animals. Ammonia is the main form of nitrogen excreted by most aquatic animals. In fishes, most of the ammonia is eliminated by the gills; the remainder enters the culture water along with the urine. Smith (1929) found that in the freshwater fishes he studied, ammonia accounted for 80 per cent of the total nitrogen excreted, with urea composing most of the remainder. According to Shirahata (1964), growing rainbow trout excrete 17 mg of ammonia nitrogen per kilogram of body weight per hour. Gerking (1955) noted that the total nitrogen excretion by bluegill sunfish was high in relation to body weight (a 29.7-g fish excreted 7.18 mg of total nitrogen per day). This, Gerking concluded, was about three and a half times as much as a warm-blooded animal would produce. That ammonia is the predominant form of nitrogen excreted by marine teleosts has been demonstrated in the sole (Delaunay, 1929), the Pacific staghorn sculpin, starry flounder, and striped seaperch (Wood, 1958).

Ammonia is also the main excretory product of aquatic invertebrates. This has been shown in many groups, including the mollusks (Delaunay, 1931; Robertson, 1954) and crustaceans (Dressel and Moyle, 1950; Binns and Peterson, 1969).

102

An excellent summary of ammonia excretion by aquatic animals has been given by Prosser and Brown (1961).

Mechanisms of Ammonia Toxicity

Although ammonia is acutely toxic to aquatic life, its degree of toxicity varies according to its chemical state. In general, the concentration of ammonia in solution (as total NH_4^+) should not be allowed to exceed 0.1 ppm. Lethal levels for many animals are even lower. However, even levels far below fatal limits may have significant effects, as will be shown below.

The dissolved oxygen level and pH are the two most important factors affecting ammonia toxicity. The latter is important because only the un-ionized form of ammonia (NH_3) appears to be poisonous to aquatic animals. Ionized ammonia (NH_4^+) is unable to pass tissue barriers (Milne et al., 1958) and thus enter an aquatic animal from the external medium.

The concentration of un-ionized ammonia on either side of a tissue barrier at a given moment depends upon the pH of the water. Usually a gradient exists at these places where the pH of the extracellular fluid (water) and the intracellular fluid (blood) are not in equilibrium. When the pH of either fluid changes, there is a shift in the concentration of un-ionized ammonia on both sides of the barrier (Warren, 1962; Warren and Schenker, 1962). This is illustrated in Fig. 27. Part a shows that the NH_3—NH_4^+ ratio is pH dependent, and that the higher the pH, the more the amount of un-ionized ammonia. Part b illustrates that only NH_3 can cross tissue barriers. In part c it can be seen that the side of the tissue barrier with the lower pH (greater concentration of H^+) attracts NH_3. Part d illustrates the fact that if the pH on both sides of the tissue barrier is changed simultaneously by the sudden infusion of free CO_2, the shift in ammonia distribution is less pronounced than when the pH on only one side of the barrier is changed by the addition of a fixed acid or base.

A reduction in the pH of water from 8 to 7 results in a tenfold decrease in the quantity of un-ionized ammonia (Downing and Merkens, 1955). The relationship between the chemical state of ammonia and the pH is given in Table 14. As the pH increases and there is a drop in the hydrogen ion concentration, an increase in the level of un-ionized ammonia occurs.

Downing and Merkens (1955) determined that trout were more resistant to ammonia at pH 7 than 8. They found that ten times as much ammonium chloride had to be added to water of pH 7 to achieve the same lethal effects at pH 8.

A decrease in dissolved oxygen increased the toxicity of un-ionized ammonia in three of the four species of freshwater fishes tested by Merkens and

Table 14. Percentage of Un-Ionized Ammonia in Solutions of Hydroxide at Three Temperatures and Various Levels of pH

pH	10 C	15 C	20 C
7.0	0.3	0.4	0.5
7.5	1.1	1.3	1.5
7.6	1.4	1.6	1.9
7.7	1.8	2.1	2.4
7.8	2.3	2.6	3.0
7.9	2.9	3.3	3.8
8.0	3.6	4.1	4.7
8.1	4.6	5.2	6.0
8.2	5.7	6.5	7.3
8.3	7.1	8.0	9.1
8.4	8.9	9.9	11.2
8.5	11.1	12.3	13.7
8.8	20.3	22.1	24.2
9.0	29.1	32.3	35.8
9.5	57.6	59.8	62.1

Derivation formula:

$$\frac{(NH_4^+)\,(OH^-)}{NH_3 \cdot H_2O} = K_b$$

K_b = dissociation constant aqueous ammonia

K_w = ionization constant H_2O

Temp.	K_b	K_w
10 C	1.570×10^{-5}	14.5346
15 C	1.652×10^{-5}	14.3463
20 C	1.710×10^{-5}	14.1669

Downing (1957). Similar findings had been reported earlier with rainbow trout (Downing and Merkens, 1955) and with minnows (Wuhrmann, 1952). Merkens and Downing (1957) discovered that increasing the dissolved oxygen from 1.5 to 8.5 ppm in the test water, at a constant temperature of 19.8 C, reduced the toxicity of un-ionized ammonia in all the concentrations they tested.

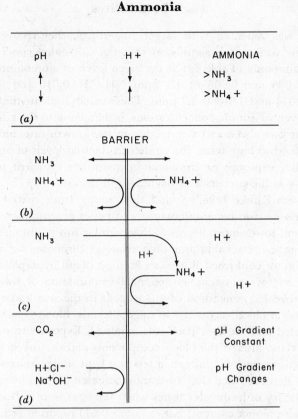

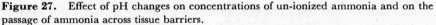

Figure 27. Effect of pH changes on concentrations of un-ionized ammonia and on the passage of ammonia across tissue barriers.

Effects of Ammonia Toxicity

In 1964 Burrows stated:

. . . The effects of even low concentrations of un-ionized ammonia are real, demonstrable, and definitely detrimental. From accumulated evidence it can be demonstrated that, when un-ionized ammonia concentrations become dominant in rearing ponds, growth rate, physical stamina, and disease resistance are impaired. . . .

The same effects, of course, occur in all culture systems.

Burrows exposed salmon fingerlings to water of pH 7.8 in which the ammonium hydroxide levels were 0.3, 0.5, and 0.7 ppm. The fish showed definite signs of hyperplasia, or "clubbing," of their gill filaments within 4 weeks. The fish did not recover after being transferred to new water. When the experiment was repeated with larger fingerlings, they recovered after 3 weeks in new water. It is significant that the calculated concentrations of un-ionized ammonia (Table 14) at the given levels of ammonium hydroxide were only 0.006 and 0.008 at 0.3 ppm NH_4OH, 0.010 and 0.012 at 0.5 ppm, and 0.014 and 0.018 at 0.7 ppm. These values indicate that un-ionized ammonia, even in minute concentrations, is damaging to the gill epithelium of fishes. Burrows also noted a reduction in the growth rate and stamina of salmon exposed to long-term, intermittent sublethal levels of ammonia. He suggested that exposure to ammonia in quantities sufficient to cause gill hyperplasia was the precursor of bacterial gill disease.

Reichenbach-Klinke (1967) found that young trout reared in ground-water with no measurable ammonia showed better growth, increased disease resistance, and lower mortality rates than similar fish maintained in stream water that gave measurable ammonia readings throughout the year. This same investigator confirmed Burrows's findings of gill hyperplasia caused by sublethal levels of ammonia. Histological examination of fish exposed to such levels revealed congestion of mucus cells in the skin and an abnormal concentration of blood corpuscles in the epidermis. Blood vessels in the liver were congested and the liver itself was inflamed. Exposure to sublethal ammonia levels also altered the blood components and destroyed erythrocytes. These conditions were magnified in test fish kept under anoxic conditions. Wolf (1957) demonstrated that ammonia increased the incidence of blue-sac disease in the fry of freshwater fishes when the eggs were cultured in water with a high ammonia content. Kawamoto (1961) reported reduced growth rates in carp caused by ammonia. During a period of 3 months, the average individual weights of the control fish increased from 2.74 to 3.13 g, while fish exposed to 0.3 ppm NH_4Cl showed an average decline in weight from 2.80 to 2.77 g.

A chronic ammonia level is the most serious problem facing the culturist. Ammonia is not species specific in effect, as are some organic metabolites; ammonia is as detrimental in mixed- as in single-species culture systems. Efficient biological filtration is the key to keeping ammonia at a minimum, although airstripping and ozonation may also help to reduce it. Susceptibility to ammonia poisoning is reduced when high turnover rates are used to maximize biological filtration and maintain the dissolved oxygen at saturation. As is evident from Table 14, a high pH (above 8) is not necessarily desirable, even in marine systems. In marine systems operated at maximum carrying capacity, an ammonia problem should be anticipated, and a stable pH between 7.5 and 8 minimizes the toxic effects.

7.3 PHEROMONES

Many animals excrete organic compounds into the environment that serve either as chemoreceptive stimulants or physiologic depressants. Karlson and Luscher (1959) named these substances *pheromones*, which they derived from the Greek *pherin*, to transfer, and *hormon*, to excite. The term has come under criticism by several authorities, among them Kirschenblatt (1962). However, pheromone seems to have prevailed and it is the term we shall use here.

In aquatic animal culture, depressant pheromones are the organic substances of major concern, although it is probable that other, less selective, organics in culture water may also be harmful. Aquatic pheromone research is only getting started and more time is needed before the exact nature and effects of depressant organic compounds—both selective and nonselective—can be categorized and measured. Present evidence indicates, however, that pheromones often have specific effects and that they are directly metabolic in origin. It is likely that many of them are also highly species specific. It was mentioned in Chapter 3 that it is impossible at this time to isolate and individually remove the many different organic substances in viable culture water. Therefore, it must simply be stated again that chemical filtration methods lower the total quantity of circulating organics; presumably the level of toxic organics, including pheromones, is lowered as well.

Most of the literature on depressant pheromones and aquatic animals is of limited value, since investigators have been content to measure the effects of these substances without determining their exact chemical structures or mechanisms of toxicity. Moreover, sublethal levels of ammonia produce similar effects and yet the possibility of ammonia acting as an interfering factor in aquatic pheromone research seems to have been universally ignored. In experiments in which the test animals were kept in closed systems without adequate biological filtration, effects such as growth inhibition and decreased fecundity could indeed have been caused by ammonia. The failure to measure ammonia invalidates all such experiments, except the ones in which a pheromone was isolated and tested, or in which the measured effects of a volume of culture water on the test animals were clearly species specific. An excellent study has been made by Yu (1968), who noted depressant effects in the zebrafish and the blue gourami (shown not to have been caused by ammonia) resulting from organic substances produced by the fishes themselves. The effects were experimentally induced by differentially extracting the substances in question and adding them to the culture water in measured amounts. Berrie and Visser (1963) isolated a compound produced by the aquatic snail, *Biomphalaria sudanica*, which proved lethal to the snails when added to their water in twice the normal concentration.[1]

Many investigators have made reference to the growth-promoting effects of old culture water. The beneficial factor involved is usually attributed to an unknown organic compound excreted into the water by the animals. This may eventually prove to be true. But on the other hand, the absence of growth-inhibiting substances could produce much the same effect as the presence of compounds that promote growth. It was shown in Chapter 1 that the conditioning process in a new culture system is slow. Many of the filter bacteria do not equilibrate with the input of their energy sources for several weeks. It was also mentioned that a new system could be conditioned more rapidly by adding a layer of filter gravel from an old system. A portion of the total bacterial population in a conditioned system is suspended and not attached to stable surfaces. It is possible, therefore, that adding old culture water simply conditions a new system faster by the infusion of viable filter organisms. The effect, of course, is the more rapid and efficient removal of growth-limiting substances and the elimination of a time lag that is often seen in the growth rates of animals kept in systems that are not yet conditioned.

---

[1] The compound isolated by Berrie and Visser was a mono-hydroxyl-tri-carboxylic acid mono-isodecyldimethyl ester with a molecular weight of 360. The empirical formula was $C_{18}H_{32}O_7$.

CHAPTER 8

Disease Prevention by
Environmental Control

There is a voluminous literature on disease treatments of aquatic animals, while comparatively little has been written on disease prevention by means of environmental control. This chapter is an attempt to reconcile the cause-versus-effect aspects of infectious disease as they relate to environmental conditions. Admittedly, any such attempt is bound to be incomplete because the relationship between disease and environmental stress is not clearly defined at the present time.

Evidence has been presented in the last three chapters that aquatic animals are acutely sensitive to slight changes in water quality. It should be emphasized again that even though most animals can control their internal responses to these changes to some extent, any response is likely to be magnified under captive conditions. In the wild the natural dispersal of both hosts and parasites allows a host species to survive even though numerous individuals may perish when other unfavorable conditions prevail. But in captivity dense crowding facilitates parasite transfer. When the water is not properly filtered and sterilized, parasites may proliferate and eventually overwhelm their hosts. Under such conditions, any natural resistance the host may have is quickly broken down by the sheer number of infectious agents.

8.1 IMMUNITY AND THE ENVIRONMENT

Infectious agents have been found in all aquatic animals, but in aquatic animal culture only transmissible forms are significant. The severity of an infection depends largely on the physiological well-being of the host. This, in turn, depends directly on environmental factors. In healthy animals infections are often *latent*—the parasites are present but not in infectious, or disease-producing, stages. Most epizootic outbreaks of disease encountered in closed-system culturing are caused by bacteria and protozoans. These organisms can normally be held in check by stringent quality control of the environment.

109

The virulence of helminth and arthropod parasites is less dependent on the water quality factors to which the host is subjected. Heavy infestations often occur on otherwise healthy animals despite carefully controlled factors like temperature, dissolved oxygen, and ammonia. The infected animals usually require treatment. Detailed discussions of fish diseases and treatments are given by Reichenbach-Klinke and Elkan (1965) and by Van Duijn (1967). Johnson (1968) gives a comprehensive survey of diseases in aquatic invertebrates, and an excellent summary of immunity and environmental stress has been given by Lom (1969).

Parasitic bacteria, protozoans and, to a lesser extent, viruses are more troublesome in closed-system culturing than are helminths and arthropods. The latter organisms can be kept out of culture systems by the methods described in the next section. However, bacteria and protozoans are usually present on the gills and external surfaces of fishes and other aquatic animals. The physiological condition of the host determines whether they remain latent or become infectious. A temporary decline in dissolved oxygen weakens the host; so does an increase in free CO_2, un-ionized ammonia, and organics. Heterotrophic bacteria proliferate as the total organic level increases and normally harmless forms may become infectious when present in large numbers. Temperature fluctuations cause considerable stress to the host and often result in chronic protozoan infections even long after the temperature has returned to normal.

Outbreaks of bacterial and protozoan infections often occur when new animals are added to a conditioned system. It is commonly thought that the infectious agents enter with the new animals and are transmitted to the established ones. But the problem may not be this simple. For instance, if the established animals are carrying the same infective organisms as the newcomers, but only in the latent stages, the addition of the new animals may cause these latent forms to become infectious. This can be brought about by several factors, including a rise in ammonia during the temporary shift in the carrying capacity of the system. There may also be a rise in dissolved organics and, in poorly buffered systems, an increase in free CO_2. In other words, water quality may still be the underlying cause even though the only visible change has been the addition of more animals. As shown in Fig. 28, a decline in water quality resulting in lowered disease resistance of the host is enough to alter the virulence of a latent parasite.

The mucus on the outer surfaces of aquatic animals serves as the first line of defense against the invasion of ectoparasites. It forms a protective sheath that helps to maintain the host–parasite balance in favor of the host. There is now substantial evidence that fish mucus contains antibodies that repel parasitic protozoans (Lom, 1969). The synthesis of antibodies by cold-blooded animals is temperature dependent. Antibody production in fish

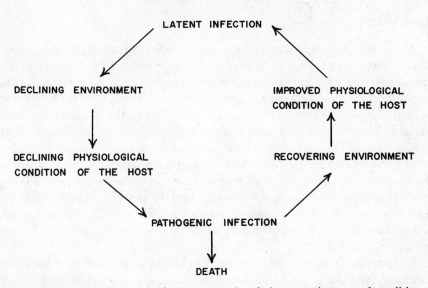

Figure 28. Virulent stages of infectious agents in relation to environmental conditions.

mucus is inhibited above or below specific temperatures, depending on the species. Lom (1969) cites evidence that raising the temperature of the culture water to 20 C caused trichodinids to drop off carp because ". . . the protective capacity of the mucus is manifested only at the elevated temperature. . . ." At 10 C the fish were unable to rid themselves of the parasites, even though the growth of the trichodinids was inhibited at the lower temperature.

Temperature fluctuations and accumulating toxic metabolites are two of the primary factors causing epizootic outbreaks of infectious bacteria and protozoans.

8.2 DISEASE PREVENTION AND THE ENVIRONMENT

Four factors are involved in environmental disease prevention: (*1*) maintaining proper environmental conditions (including water quality and stable temperatures), (*2*) sterilizing the circulating culture water, (*3*) providing adequate nutrition, and (*4*) preventing the introduction of infectious agents from outside sources. The first two factors have been discussed in previous chapters, and the third is not within the scope of this book. The fourth factor will now be considered.

Raw Water

Many infectious agents are present in unprocessed natural, or raw, water. Raw water should never be pumped directly from a natural source into a culture system. Makeup for partial water changes should be prefiltered with rapid sand filters to remove the initial turbidity. Then it should be filtered with diatomaceous earth. DE filters are very effective in removing any bacteria and protozoans that may have passed through the sand beds. After filtration with DE, the water should be kept in darkened storage vats for 2 weeks and moderately aerated. Most parasites die within this time when no hosts are available. This means that at least two vats should be available, each holding enough water for the standard 10 per cent change routinely provided for each culture system biweekly. There should also be enough extra volume in each vat to start up a new culture system in case of emergency.

In closed-system culturing, prefiltration of raw water is carried out intermittently and the filters are not in continuous operation. The prefiltration system (rapid sand and DE filters in series) should be designed so that the filters can be recycled independently of the storage vats and culture systems. This arrangement is shown diagrammatically by dotted lines in Fig. 29. The position of the sterilizer (ozone or uv irradiation) is also shown in the figure. The water should be sterilized as it leaves the vats on its way to the culture systems after the 2-week aging period.

The prefiltration system should be chemically sterilized immediately after processing a batch of raw water, then drained and left dry until the next batch is filtered. Sterilization with uv irradiation or ozone is not effective in this case, since neither one kills infectious organisms deep in the sand beds or attached to the DE filter elements, walls, and bays. Superchlorination is the most effective technique, although great care should be taken to remove all traces of residual chlorine.

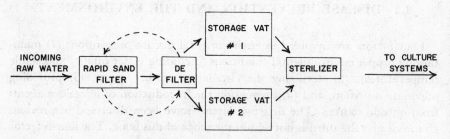

Figure 29. Prefiltering arrangement for processing large volumes of natural water.

1. Adjust the valves so that the prefilters operate as a closed system (recycle).

2. Backwash the DE filter. Do not precoat.

3. Add sodium hypochlorite until the level of free chlorine is at 50 ppm as determined by the OTO flash test (see American Public Health Association et al., 1965).

4. Turn on the pumps and recycle the chlorinated water through the filters for 2 hr.

5. Reduce the free chlorine to zero with sodium thiosulfate ($Na_2S_2O_3$). The amount required will vary and must be determined by trial and error.

6. Drain the filters and leave them dry until the next batch of raw water is prefiltered.

7. When starting prefiltration again, let the system flush to waste for 30 min before diverting the water to the storage vats.

Infected Animals

New stocks of animals should not be added to the main systems until they have been isolated (quarantined) and have proved to be free of infections (microorganisms) or infestations (macroorganisms) for 4 weeks.

Live Food

Live food may be unsafe unless it is tank raised. Cultured brine shrimp are suitable to feed to both freshwater and marine animals. Forms of live food collected in the wild are potential carriers of parasites. Live food collected from marine waters should not be fed to marine animals; freshwater food organisms should not be fed to freshwater animals unless they are tank raised. However, parasites normally present on marine food organisms can seldom survive immersion in fresh water and are safe to feed to freshwater animals. The reverse is also true, and freshwater forms, such as wild *Daphnia*, are safe to feed to marine animals. Freshwater minnows should never be fed to carnivorous freshwater fishes, although they are safe to feed to marine species.

In all instances, commercially prepared foods are preferable to live or frozen animal flesh from the standpoint of disease transfer.

8.3 TREATING DISEASES

Without the proper controls, treating diseases and at the same time maintaining stable environmental conditions are antagonistic processes. The culturist who is too quick with his beaker of remedies may aggravate an already serious situation by killing beneficial microbes in the filter bed. When this happens, the decline in water quality accelerates and still more factors combine to plague the animals. With this in mind, any treatment program must be approached thoughtfully and with the realization that failure is imminent in a deteriorating environment. In unskilled hands, the "cure" may be more lethal than the disease.

In most cases it is best not to treat animals infected by bacteria or protozoans. Loss of stocks to these organisms, as previously mentioned, can often be traced to an environmental factor. The most heavily infected stocks must be removed and the factor responsible corrected. Animals with mild infections and in otherwise good health stand a good chance of recovery.

Helminth and arthropod parasites have greater resistance to the immune responses of the host and cannot be handled in the same manner. The only sure way to eradicate these organisms is to remove all the animals and superchlorinate the culture water, tank, and filtration system. A free chlorine level of 50 ppm sustained for 2 hr is sufficient. The chlorine level must then be reduced to zero with sodium thiosulfate and the system drained and flushed thoroughly with tap water. The system can then be refilled with newly processed culture water and conditioned again with new stocks that have been quarantined and acclimated. This technique is admittedly drastic, but it is also the only one that is completely effective. Heavy infestations of helminths and arthropods are seldom a problem if quality control procedures, like the ones outlined in the previous section, are stringently followed.

The actual selection of medication has been the subject of several books, so nothing specific will be mentioned here. In selecting a compound, however, three things should be kept in mind. First, one should be chosen that is as specific as possible for the infectious agent in question. Second, compounds that cause aquatic animals to shed copious quantities of mucus should be avoided. When the mucus is destroyed the animals are left unprotected against secondary invasion by other pathogens, or even against reinfection. Heavy metals, therefore, are not recommended. Third, diseased animals should never be treated directly in the culture system. Many substances, such as antibiotics and formalin, interrupt nitrification. For this reason, treatment tanks equipped with biological filtration are seldom functional. Ammonia conversion in these tanks should be carried out chemically by air-stripping or ozonation.

CHAPTER 9

Laboratory Tests[1]

Ammonia, nitrite, and nitrate are measured spectrophotometrically by extinction ($=$absorbance, or E) or by per cent transmittance (T) on a Bausch & Lomb Spectronic 20 or its equivalent. More precise instrumentation is unnecessary in routine culture water management.

The values obtained from the spectrophotometer are converted to parts per million by comparing them against standard calibration curves that have been plotted on log paper with known concentrations of ammonia and nitrite. For example, solutions are prepared containing 0, 0.2, 0.4, and 0.6 ppm. Solutions of 0.366 g $(NH)_2SO_4$/liter and 0.150 g $NaNO_2$/liter contain 100 ppm NH_4^+ and NO_2^-, respectively. Each solution is diluted to the above concentrations and the points are plotted and connected. Nitrite and nitrate are read from the same curve.

The accuracy of these tests is affected by salinity within the range 25–38‰. For ammonia, readings obtained will be 0.62 times (in extinction) the values obtained from the same concentration of ammonia in distilled water. When using per cent transmittance the conversion factor is $0.62(100-T)$. If the conversions are made when the points are plotted during calibration, the values from sea water samples can be read directly from the curves.

When reading nitrate, the conversion is 0.85 times (in extinction) the values obtained from the same concentration of nitrate in distilled water. With per cent transmittance, $0.85(100-T)$ is used.

Reagent blanks made up from distilled water should be run before each series of tests and the values of the samples adjusted accordingly.

9.1 AMMONIA (as total NH_4^+)

The following test is for total ammonia. The temperature and pH of the sample should also be measured and the per cent of un-ionized ammonia estimated from Table 14.

[1] The ammonia, nitrite, and nitrate tests, as given here, have been adapted from two sources: (1) *A Practical Handbook of Seawater Analysis*, J. D. H. Strickland and T. R. Parsons (1968), Bull. 167, reprinted by permission of the Fisheries Research Board of Canada, Ottawa, Canada, copyright 1968; and (2) *Aquarium Water Chemistry Manual*, reprinted by permission of Aquarium Systems, Inc., Eastlake, Ohio, copyright 1968.

115

Reagents

Acetate buffer	Dissolve 11 g sodium acetate trihydrate in a little distilled water. Add 65 ml glacial acetic acid and distilled water to 500 ml. Maximum life: 6 months.
Chloramine-T solution	Dissolve 2.5 g chloramine-T in 50 ml distilled water. Maximum life: 1 week.
Mono reagent	Dissolve 5 g 3-methyl-1-phenyl-2-pyrazolin-5-one (Eastman) in 2 liters distilled water with heat. Reagent must age for 2 weeks before use. Maximum life: 1 year.
Pyrazolone reagent	Dissolve 80 mg Bis reagent (3,3-dimethyl-1, 1-diphenyl-4,4-bi-2-pyrazoline-5,5-dione-- (Eastman) in 80 ml pyridine and add 400 ml "mono" reagent. Maximum life: 2 hr.
Carbon tetrachloride	Reagent grade.

Procedure

1. To an 80-ml sample add 1 ml acetate buffer and 1 ml chloramine-T.
2. After 90 sec add 40 ml pyrazolone reagent.
3. After 60 sec add 35 ml carbon tetrachloride and shake for 1 min in a separatory funnel.
4. Allow to settle and filter off some of the carbon tetrachloride through Whatman #1 filter paper into a cuvet.
5. Wipe the cuvet clean of fingerprints and read against pure carbon tet at 450 mμ.

9.2 NITRITE AND NITRATE

Reagents

Acetone	Reagent grade.
Buffer	Mix 20 ml sodium hydroxide solution with 20 ml phenol solution.
Copper sulfate solution	Dissolve 100 mg copper sulfate (hydrous) in 1 liter distilled water.
Dihydrochloride solution	Dissolve 100 mg of N-(1-naphthyl)-ethylene-diamine dihydrochloride (MCB 6688) in 100 ml distilled water. Discard when it turns brown.
Hydrazine sulfate	Dissolve 3.625 g hydrazine sulfate in 500 ml distilled water. Maximum life: 2 months.

Phenol solution Dissolve 9.072 g phenol in 200 ml distilled water. Maximum life: 2 months.

Reducer Mix 20 ml copper sulfate solution with 20 ml hydrazine sulfate.

Sodium hydroxide solution Dissolve 1 oz (28.35 g) sodium hydroxide in distilled water. Cool and dilute to 2 liters. Maximum life: 6 months.

Sulfanilamide solution Dissolve 5 g sulfanilamide (Mall 8393) in a mixture of 50 ml concentrated hydrochloric acid and 300 ml distilled water. Bring to 500 ml with distilled water.

Procedure (Nitrite)

1. To a 50-ml sample add 4 ml sulfanilamide solution.
2. After 2 min add 4 ml dihydrochloride solution.
3. After 10 min to 2 hr read against distilled water at 543 mμ.

Procedure (Nitrate)

1. Take 0.5 ml test water and dilute it with 50 ml distilled water. This is the 50-ml sample.

2. To the 50-ml sample add 2 ml buffer and 1 ml reducer. Place sample in a constant temperature bath between 70 and 80 F for 20 hr. Shield from light.

3. Remove sample from bath and add 2 ml acetone. After 2 min proceed as in nitrite test, but multiply the result by 100.

9.3 MEASUREMENT OF SALINITY AND SPECIFIC GRAVITY

Precise determination of salinity can be made by measuring the electrical conductivity or freezing point depression. Salinity can also be calculated from the chlorinity, as shown by equation 16 on p. 85. In routine culturing, such precise methods are rarely necessary, and direct measurements with hydrometers are accurate enough.

Hydrometers can be purchased with scales that measure the density of water either directly as specific gravity, or indirectly after conversion to salinity. Specific gravity is preferable. Only hydrometers that have been calibrated against NBS standards by the manufacturer should be used.

9.4 DISSOLVED OXYGEN

Dissolved oxygen in fresh, brackish, or sea water can be measured either in the laboratory by the titration method, or *in situ* with a portable electronic meter.

Titration Method[2]

Reagents

Manganous sulfate solution	Dissolve any one of the following manganous sulfates (hydrous) in a little distilled water, then filter and dilute to 100 ml: 48 g $MgSO_4 \cdot 4H_2O$; 40 g $MgSO_4 \cdot 2H_2O$; or 36.4 g $MgSO_4 \cdot H_2O$.
Alkali-iodide-azide	Dissolve 50 g NaOH (or 70 g KOH) and 15 g KI in distilled water and dilute to 100 ml. Add 1 g NaN_3 dissolved in 4 ml distilled water.
Starch solution	Blend 0.5–0.6 g soluble starch in a beaker with a little distilled water. Add to a beaker containing 100 ml boiling water and continue boiling for a few minutes. Add a few drops of toluene to preserve.
Sodium thiosulfate stock solution	Dissolve 2.48 g $Na_2S_2O_3 \cdot 5H_2O$ in boiled and cooled distilled water. Dilute to 100 ml with distilled water and add 0.5 ml chloroform to preserve.
Standard sodium thiosulfate titrant (0.025N)	Dilute 25 ml sodium thiosulfate stock solution to 100 ml with distilled water. Add 0.5 ml chloroform to perserve. 1 ml titrant = 0.20 mg dissolved oxygen (DO) per 1.00 ml.
Sulfuric acid	Concentrated (36N).
Potassium dichromate standard solution (0.025N)	Dry some $K_2Cr_2O_7$ at 103 C for 2 hr, then weigh out 0.0225 g and dissolve in 100 ml distilled water in a volumetric flask.

Standardization

1. Dissolve 2 g KI in 100–150 ml distilled water. Add 10 ml of a solution containing 1 part sulfuric acid and 9 parts distilled water.

2. Add 20 ml potassium dichromate stock solution.

3. Dilute to 400 ml with distilled water and titrate with the thiosulfate titrant. Add starch (1–2 ml) near the end of the titration when a pale yellow

---

[2] Adapted from *Standard Methods for the Examination of Water and Wastewater*, by the American Public Health Association et al. (1965), 12th ed., and reprinted by permission of the American Public Health Association, New York, New York, copyright 1965.

color is reached. 20 ml titrant is required when the thiosulfate titrant solution is exactly 0.025N.

Test Procedure

1. Collect samples in 300-ml glass-stoppered bottles. Siphon water into the bottles keeping the discharge end of the siphon tube completely submerged. Allow water from the culture system to overflow 3–4 times the volume of the bottle. Insert the stopper *without trapping air*.

2. Add 2 ml manganous sulfate solution, followed by 2 ml alkali-iodide-azide solution. Keep the end of the pipet submerged in the water sample.

3. Replace the stopper without trapping air. Invert the bottle several times, allowing the precipitate to settle halfway each time before inverting again.

Note: In brackish- and seawater samples wait 10 min before proceeding to the next step.

4. Add 2 ml sulfuric acid to the sample by letting it run out of the pipet down the inside of the bottle neck. Restopper and invert gently several times until the precipitate dissolves.

5. Pour 200 ml of the sample into a beaker. Titrate with thiosulfate titrant to a pale yellow color. Add 1–2 ml starch solution and continue titration until the blue color disappears. After the starch has been added, avoid going past the end point by adding the titrant a drop at a time and gently swirling the beaker after each drop. 1 ml titrant = 1 ppm dissolved oxygen (DO).

Note: If the amount of titrant used during standardization is not exactly 20 ml (step 3), then at the end of the test procedure the amount of DO in the sample can be determined as follows:

$$\text{ppm DO} = \frac{\text{ml test titrant} \times 10}{\text{actual ml standardization titrant}}$$

Credits for Illustrative Material

CHAPTER 1

Figure 1   R. Y. Stanier, M. Doudoroff, and E. A. Adelberg, *The Microbial World*, 2nd ed., copyright 1957, 1963. Adapted and reprinted by permission of Prentice-Hall, Inc., Englewood Cliffs, New Jersey.

Figure 2   Adapted from A. Kawai, Y. Yoshida, and M. Kinata (1964). Reprinted by permission of A. Kawai, Kyoto University, Kyoto, Japan.

Table 1   Courtesy Gardner-Denver, Quincy, Illinois.

CHAPTER 2

Figure 6   Courtesy Paragon Swimming Pool Company, Inc., Pleasantville, New York.

Figure 7   Courtesy BIF, a unit of General Signal Corp., Providence, Rhode Island.

Figure 8   Courtesy Keene Corp., Fluid Handling Div., Cookeville, Tennessee.

Figure 9   Adapted by courtesy of Johns-Manville Products Corp., New York, New York.

Figure 10   Adapted by courtesy of T. Shriver & Company, Inc., Harrison, New Jersey.

Figure 11   Courtesy Johns-Manville Products Corp.

Figure 12   Adapted by courtesy of Johns-Manville Products Corp.

Figure 13   Courtesy Johns-Manville Products Corp.

CHAPTER 3

Figure 14   Adapted by courtesy of Barnebey-Cheney, Columbus, Ohio.

Table 2   Adapted from J. D. Parkhurst, F. D. Dryden, G. N. McDermott, and J. English, "Pomona Activated Carbon Pilot Plant," *J. Water Pollution Control Federation*, **39**, R70 (1967).

Figure 18   Adapted from J. D. Parkhurst, F. D. Dryden, G. N. McDermott, and J. English, "Pomona Activated Carbon Pilot Plant," *J. Water Pollution Control Federation*, **39**, R70 (1967).

Figure 19   From R. Eliassen, B. M. Wyckoff, and D. C. Tonkin (1965). Reprinted from the *Journal of the American Water Works Association*, copyright 1965 by the American Water Works Association, Inc., 2 Park Avenue, New York, New York 10016.

Figure 20   From R. Eliassen, B. M. Wyckoff, and D. C. Tonkin (1965). Reprinted from the *Journal of the American Water Works Association*, copyright 1965 by the American Water Works Association, Inc., 2 Park Avenue, New York, New York 10016.

Figure 21   From R. Eliassen, B. M. Wyckoff, and D. C. Tonkin (1965). Reprinted from the *Journal of the American Water Works Association*, copyright 1965 by the American Water Works Association, Inc., 2 Park Avenue, New York, New York 10016.

Table 3   Adapted from E. Rubin, R. Everett, Jr., J. J. Weinstock, and H. M. Schoen (1963). Contaminant removal from sewage plate effluents by foaming. PHS Publ. No. 999-WF-5. 56 pp.

Figure 22   Adapted from E. Sander (1967). Reprinted by permission of E. Sander, 3151 Eltze, Am Osterberg, West Germany.

Figure 23   Adapted from E. Sander (1967). Reprinted by permission of E. Sander, 3151 Eltze, Am Osterberg, West Germany.

CHAPTER 4

Table 4   Adapted from J. P. Riley and G. Skirrow, Eds. *Chemical Oceanography*, Vol. 1, copyright 1965 by Academic Press, Inc. Reprinted by permission of Academic Press, Inc., New York, New York.

Table 5   Adapted from J. P. Riley and G. Skirrow, Eds. *Chemical Oceanography*, Vol. 1 copyright 1965 by Academic Press, Inc. Reprinted by permission of Academic Press, Inc., New York, New York.

CHAPTER 5

Table 6   Calculated from data given in C. N. Murray and J. P. Riley (1969).

CHAPTER 6

Table 7   Courtesy G. M. Mfg. and Instrument Corp., Inc., New York, New York.

Table 8   Adapted from H. J. M. Bowen. *Trace Elements in Biochemistry*, copyright 1966 by Academic Press, Inc. Reprinted by permission of Academic Press, Inc., New York, New York.

Table 9   Courtesy Aquarium Systems, Inc., Eastlake, Ohio.

Figure 26   Adapted and reprinted by permission of International Salt Company, Clarks Summit, Pennsylvania.

CHAPTER 7

Figure 27   From K. S. Warren and S. Schenker (1962). Reprinted from the *American Journal of Physiology*, copyright 1962 by the American Physiological Society, Bethesda, Maryland.

Literature Cited

American Public Health Association, American Water Works Association, Water Pollution Control Federation
    1965   Standard methods for the examination of water and wastewater. 12th ed. American Public Health Association, New York. 769 pp.

Aquarium Systems, Inc.
    1968   Aquarium water chemistry manual. Aquarium Systems, Eastlake, Ohio. 16 pp.

Arnon, D. I., and P. R. Stout
    1939   The essentiality of certain elements in minute quantity for plants, with special reference to copper. *Plant Physiol.*, **14**: 371–375.

Atz, J. W.
    1964a   Some principles and practices of water management for marine aquariums. *In* Sea-water systems for experimental aquariums: a collection of papers, J. R. Clark and R. L. Clark (eds.). U. S. Dept. of the Interior, Bureau of Sport Fisheries and Wildlife. Res. Rept. 63. 192 pp.

Atz, J. W.
    1964b   Self-inhibition by captive fishes through the water in which they live. *Aquasphere* (Boston), **2**: 11–13.

Barber, R. T.
    1966   Interaction of bubbles and bacteria in the formation of organic aggregates in sea water. *Nature*, **211**: 257–258.

Basu, S. P.
    1959   Active respiration of fish in relation to ambient concentrations of oxygen and carbon dioxide. *J. Fish. Res. Board Can.*, **16**: 175–212.

Baylor, E. R., and W. H. Sutcliffe
    1963   Dissolved organic matter in seawater as a source of particulate food. *Limnol. Oceanog.*, **8**: 369–371.

Bean, E. L.
    1959   Ozone production and costs. *In* Ozone chemistry and technology. *Advan. in Chem. Ser.*, **21**: 465 pp.

Bedford, R. H.
    1933   Marine bacteria of the northern Pacific Ocean. The temperature range of growth. *Contrib. Can. Biol. Fish.*, **7**: 433–438.

Benoit, R. F., and N. A. Matlin
    1966   Control of *Saprolegnia* on eggs of rainbow trout (*Salmo gairdnerii*) with ozone. *Trans. Amer. Fish. Soc.*, **95**: 430–432.

Berner, R. A.
    1966   Diagenesis of carbonate sediments: interaction of magnesium in sea water with mineral grains. *Science*, **153**: 188–191.

Berner, R. A.
    1968   Calcium carbonate concretions formed by the decomposition of organic matter. *Science*, **159**: 195–197.

Literature Cited

Berrie, A. D., and S. A. Visser
  1963 Investigations of a growth-inhibiting substance affecting a natural population of freshwater snails. *Physiol. Zool.*, **36**: 167–173.
Binns, R., and A. J. Peterson
  1969 Nitrogen excretion by the spiny lobster *Jasus edwardsi* (Hutton): the role of the antennal gland. *Biol. Bull.*, **136**: 147–153.
Black, E. C., F. E. J. Fry, and V. S. Black
  1954 The influence of carbon dioxide on the utilization of oxygen by some freshwater fish. *Can. J. Zool.*, **32**: 408–420.
Bowen, H. J. M.
  1966 Trace elements in biochemistry. Academic Press, New York. 241 pp.
Burrows, R. E.
  1964 Effects of accumulated excretory products on hatchery-reared salmonids. U.S. Dept. of the Interior, Bureau of Sport Fisheries and Wildlife. Res. Rept. 66. 12 pp.
Burrows, R. E., and B. D. Combs
  1968 Controlled environments for salmon propagation. *Progressive Fish-Culturist*, **30**: 123–136.
Carlucci, A. F., and P. M. Williams
  1965 Concentration of bacteria from sea water by bubble scavenging. *J. Cons. Perm. Int. Explor. Mer.*, **30**: 28–33.
Carpenter, K. E.
  1927 The lethal action of soluble metallic salts on fishes. *Brit. J. Exp. Biol.*, **4**: 378–390.
Carpenter, K. E.
  1930 Further researches on the action of metallic salts on fishes. *J. Exp. Zool.*, **56**: 407–422.
Chave, K. E.
  1965 Calcium carbonate: association with organic matter in surface seawater. *Science*, **148**: 1723–1724.
Chave, K. E., and E. Suess
  1967 Suspended minerals in seawater. *Trans. N. Y. Acad. Sci. (Ser. II)*, **29**: 991–1000.
Clesceri, N. L.
  1968 Physical and chemical removal of nutrients. *In* Algae, man, and the environment, D. F. Jackson (ed). Syracuse University Press, Syracuse. 554 pp.
Culp, G., and A. Slechta.
  1966 Nitrogen removal from waste effluents. *Public Works*, **97**: 90–91.
Dahlberg, M. L., D. L. Shumway, and P. Doudoroff
  1968 Influence of dissolved oxygen and carbon dioxide on swimming performance of largemouth bass and coho salmon. *J. Fish. Res. Board Can.*, **25**: 49–70.
Deguchi, Y.
  1960 Contribution to the study of aquarium conditions. II. Dissolved organic matter in aquarium water. *Bull. Res. Coll. Agr. Vet. Sci. (Nippon Univ.)*, **11**: 109–125.
Delaunay, H.
  1929 Sur l'excrétion azotée des poissons. *Compt. rend. Soc. Biol.* (Paris), **101**: 371–372.
Delaunay, H.
  1931 L'excrétion azotée des invertébrés. *Biol. Rev.*, **6**: 265–301.

Dickerman, J. M., A. O. Castraberti, and J. E. Fuller
  1954   Action of ozone on water-borne bacteria. *J. New Engl. Water Works Assoc.*, **68**: 11–14.
Doudoroff, P.
  1952   Some recent developments in the study of toxic industrial wastes. *Proc. Fourth Pacific Northwest Ind. Waste Conf.*, State Coll. of Washington, Pullman, Wash., pp. 21–25.
Doudoroff, P., and M. Katz
  1953   Critical review of literature on the toxicity of industrial wastes and their components to fish. II. The metals as salts. *Sewage Ind. Wastes*, **25**: 802–839.
Downing, K. M., and J. C. Merkens
  1955   The influence of dissolved-oxygen concentration on the toxicity of un-ionized ammonia to rainbow trout (*Salmo gairdnerii* Richardson). *Ann. Appl. Biol.*, **43**: 243–246.
Dresel, E. B., and V. Moyle
  1950   Nitrogen excretion by amphipods and isopods. *J. Exp. Biol.*, **27**: 210–225.
Duijn, C. van, Jr.
  1967   Diseases of fish. 2nd ed. Iliffe Books, London. 309 pp.
Eliassen, R., B. M. Wyckoff, and D. C. Tonkin
  1965   Ion exchange for reclamation of reusable supplies. *J. Amer. Water Works Assoc.*, **57**: 1113–1122.
Ellis, M. M.
  1937   Detection and measurement of stream pollution. Bull. No. 22, U.S. Bureau of Fisheries. *Bull. Bur. Fish.*, **48**: 365–437.
Evans, R. M., F. C. Purdie, and C. P. Hickman, Jr.
  1962   The effect of temperature and photoperiod on the respiratory metabolism of rainbow trout (*Salmo gairdnerii*). *Can. J. Zool.*, **40**: 107–118.
Fair, G. M., and J. C. Geyer
  1958   Elements of water supply and waste-water disposal. Wiley, New York. 615 pp.
Fetner, R. H., and R. S. Ingols
  1959   Bactericidal activity of ozone and chlorine against *Escherichia coli* at 1 degree C. *In* Ozone chemistry and technology. *Advan. Chem. Ser.*, **21**: 465 pp.
Fitzgerald, G. P.
  1963   Factors affecting the toxicity of copper to algae and fish. Proc. Amer. Chem. Soc. Meeting, New York. pp. 21–24.
Fontaine, A. R., and F. S. Chia
  1968   Echinoderms: an autoradiographic study of assimilation of dissolved organic molecules. *Science*, **161**: 1153–1155.
Frisch, N. W., and R. Kunin
  1960   Organic fouling of anion-exchange resins. *J. Amer. Water Works Assoc.*, **52**: 875–887.
Fry, F. E. J., and J. S. Hart
  1948   The relation of temperature to oxygen consumption in the goldfish. *Biol. Bull.*, **94**: 66–77.
Garrett, W. D.
  1967   Stabilization of air bubbles at the air–sea interface by surface-active material. *Deep-Sea Res.*, **14**: 661–672.
Gerking, S. D.
  1955   Endogenous nitrogen excretion of bluegill sunfish. *Physiol. Zool.*, **28**: 283–289.

Herald, E. S., R. P. Dempster, C. Wolters, and M. L. Hunt
    1962   Filtration and ultraviolet sterilization of seawater in large closed, and semi-closed aquarium systems. *Bull. Inst. Oceanog.*, Num. Spéc., **1B**: 49–61.

Herbert, D. W. M., and J. M. VanDyke
    1964   The toxicity to fish of mixtures of poisons. I. Copper–ammonia and zinc–phenol mixtures. *Ann. Appl. Biol.*, **53**: 415–421.

Hirayama, K.
    1965   Studies on water control by filtration through sand bed in a marine aquarium with closed circulating system. I. Oxygen consumption during filtration as an index in evaluating the degree of purification of breeding water. *Bull. Jap. Soc. Sci. Fish.*, **31**: 977–982.

Hirayama, K.
    1966   Studies on water control by filtration through sand bed in a marine aquarium with closed circulating system. IV. Rate of pollution of water by fish, and the possible number and weight of fish kept in an aquarium. *Bull. Jap. Soc. Sci. Fish.*, **32**: 20–26.

Hoar, W. S., and G. B. Robertson
    1959   Temperature resistance of goldfish maintained under controlled photoperiods. *Can. J. Zool.*, **37**: 419–428.

Hoff, J. G., and J. R. Westman
    1966   The temperature tolerances of three species of marine fishes. *J. Mar. Res.*, **24**: 131–140.

Honig, C.
    1934   Nitrates in aquarium water. *J. Mar. Biol. Assoc. U.K.*, **19**: 723–725.

Jansen, J. F., and Y. Kitano
    1963   The resistance of recent marine carbonate sediments to solution. *J. Oceanog. Soc. Japan*, **18**: 208–219.

Johnson, P. T.
    1968   An annotated bibliography of pathology in invertebrates other than insects. Burgess Publishing Company, Minneapolis. 322 pp.

Jones, J. R. E.
    1938   The relative toxicity of salts of lead, zinc, and copper to the stickleback (*Gasterosteus aculeatus* L.) and the effect of calcium on the toxicity of lead and zinc salts. *J. Exp. Biol.*, **15**: 394–407.

Jones, J. R. E.
    1939   The relation between the electrolytic solution pressures of the metals and their toxicity to the stickleback (*Gasterosteus aculeatus* L.). *J. Exp. Biol.*, **16**: 425–437.

Kanungo, M. S., and C. L. Prosser
    1959   Physiological and biochemical adaptation of goldfish to cold and warm temperature. I. Standard and active oxygen consumptions of cold- and warm-acclimated goldfish at various temperatures. *J. Cell. Physiol.*, **54**: 259–264.

Karlson, P., and M. Luscher
    1959   "Pheromones": a new term for a class of biologically active substances. *Nature*, **183**: 55–56.

Kawai, A., Y. Yoshida, and M. Kinata
    1964   Biochemical studies on the bacteria in aquarium with circulating system. I. Changes of the qualities of breeding water and bacterial population of the aquarium during fish cultivation. *Bull. Jap. Soc. Sci. Fish.*, **30**: 55–62.

Kawai, A., Y. Yoshida, and M. Kinata
  1965   Biochemical studies on the bacteria in aquarium with circulating system. II.
         Nitrifying activity of the filter sand. *Bull. Jap. Soc. Sci. Fish.*, **31**: 65–71.
Kawamoto, N. Y.
  1961   The influence of excretory substances of fish on their own growth. *Progressive
         Fish-Culturist*, **23**: 70–75.
Kern, D. M.
  1960   The hydration of carbon dioxide. *J. Chem. Educ.*, **37**: 14–23.
Kirschenblatt, J.
  1962   Termonology of some biologically active substances and validity of the term
         "pheromone." *Nature*, **195**: 916–917.
Kuhl, H., and H. Mann
  1962   Modellversuche zum Stoffhaushalt in Aquarien bei verschiedenem Salzgehalt.
         Kiel. Meeresforsch., **18**: 89–92.
Kuhn, P. A.
  1956   Removal of ammonia nitrogen from sewage effluent. M. S. thesis. Dept. of
         Civil Engineering, University of Wisconsin.
Kunin, R.
  1963   Helpful hints in ion exchange technology. *Reprinted from* Amber-Hi-Lites,
         1961 and 1962. Rohm and Haas Company, Philadelphia. 12 pp.
Lebout, H.
  1959   Fifty years of ozonation at Nice. *In* ozone chemistry and technology. *Advan.
         Chem. Ser.*, **21**: 465 pp.
Lees, H.
  1952   The ammonia-oxidizing systems of *Nitrosomonas. Biochem. J.*, **52**: 134–139.
Lenfant, C., and K. Johansen
  1966   Respiratory function in the elasmobranch *Squalus suckleyi* G. *Respir. Physiol.*,
         1: 13–29.
Lloyd, R., and D. W. M. Herbert
  1962   The effect of the environment on the toxicity of poisons to fish. *J. Inst. Publ.
         Health Eng.*, July: 132–143.
Lom, J.
  1969   Cold-blooded vertebrate immunity to protozoa. *In* Immunity to parasitic
         animals, Vol. 1, G. J. Jackson, R. Herman, and I. Singer (eds.). Appleton-
         Century-Crofts, New York. 292 pp.
Martinez, W. W.
  1962   Phosphate and nitrate removal from treated sewage by exchange resins.
         M. S. thesis. Dept. of Civil Engineering, Pennsylvania State University.
Merkins, J. C., and K. M. Downing
  1957   The effect of tension of dissolved ogygen on the toxicity of un-ionized ammonia
         to several species of fish. *Ann. Appl. Biol.*, **45**: 521–527.
Meuwis, A. L., and M. J. Heuts
  1957   Temperature dependence of breathing rate in carp. *Biol. Bull.*, **112**: 97–107.
Milne, M. D., B. H. Scribner, and M. A. Crawford
  1958   Non-ionic diffusion and the excretion of weak acids and bases. *Amer. J. Med.*,
         24: 709–729.
Miyake, Y., and T. Abe
  1948   A study on the foaming of sea water. *J. Mar. Res.*, **7**: 67–73.
Moore, H. B.
  1958   Marine ecology. Wiley, New York. 493 pp.

128   Literature Cited

Morris, J. C., and W. J. Weber, Jr.
     1964   Adsorption of biochemically resistant material from solution. I. PHS Publ. No. 999-WF-11. 74 pp.

Morris, R. W.
     1962   Body size and temperature sensitivity in the cichlid fish, *Aqeuidens portalegrensis* (Hensel). *Amer. Naturalist*, **96**: 35–50.

Morris, R. W.
     1965   Thermal acclimation of metabolism of the yellow bullhead, *Ictalurus natalis* (Le Sueur). *Physiol. Zool.*, **38**: 219–227.

Murray, C. N., and J. P. Riley
     1969   The solubility of gases in distilled water and in sea water. II. Oxygen. *Deep-Sea Res.*, **16**: 311–320.

Nesselson, R. J.
     1954   Removal of inorganic nitrogen from sewage effluent. Ph.D. thesis. Dept. of Civil Engineering, University of Wisconsin.

Nigrelli, R. F., and G. D. Ruggieri, S. J.
     1966   Enzootics in the New York Aquarium caused by *Cryptocaryon irritans* Brown, 1951 (= *Ichthyophthirius marinus* Sikama, 1961), a histophagous ciliate in the skin, eyes and gills of marine fishes. *Zoologica*, **51**: 97–102.

Parkhurst, J. D., F. D. Dryden, G. N. McDermott, and J. English
     1967   Pomona 0.3 MGD activated carbon pilot plant. *J. Water Pollution Control Federation*, **39**: R 70–R 81.

Potts, W. T. W., and G. Parry
     1965   Osmotic and ionic regulation in animals. Pergamon Press, London. 432 pp.

Prosser, C. L., L. M. Barr, R. D. Pinc, and C. Y. Lauer
     1957   Acclimation of gold fish to low concentrations of oxygen. *Physiol. Zool.*, **39**: 137–141.

Prosser, C. L., and F. A. Brown
     1961   Comparative animal physiology. 2nd ed. W. B. Saunders, Philadelphia. 688 pp.

Quastel, J. H., and P. G. Scholefield
     1951   Biochemistry of nitrification in soil. *Bacteriol. Rev.*, **15**: 1–53.

Rahn, H.
     1966   Aquatic gas exchange: theory. *Respir. Physiol.*, **1**: 1–12.

Reichenbach-Klinke, H. H.
     1967   Untersuchungen uber die Einwirkung des Ammoniakgehalts auf den Fischorganismus. *Arch. Fischereiwissenschaft*, **17**: 122–132.

Reichenbach-Klinke, H. H., and E. Elkan
     1965   The principal diseases of lower vertebrates. Academic Press, New York. 600 pp.

Riley, G. A.
     1963   Organic aggregates in seawater and the dynamics of their formation and utilization. *Limnol. Oceanog.*, **8**: 372–381.

Riley, J. P., and G. Skirrow, eds.
     1965   Chemical oceanography, Vol. 1. Academic Press, New York. 712 pp.

Robertson, J. D.
     1954   The chemical composition of the blood of some aquatic chordates, including members of the Tunicata, Cyclostomata and Osteichthyes. *J. Exp. Biol.*, **31**: 424–442.

Roeder, M., and R. H. Roeder
     1964   The respriation of a graded series of two species of small whole xiphophorin fishes. *J. Cell. Physiol.*, **63**: 115–117.

Rubin, E., R. Everett, Jr., J. J. Weinstock, and H. M. Schoen
  1963 Contaminant removal from sewage plant effluents by foaming. PHS Publ. No. 999-WF-5. 56 pp.

Saeki, A.
  1958 Studies on fish culture in filtered closed-circulation aquaria. I. Fundamental theory and system design standards. *Bull. Jap. Soc. Sci. Fish.*, **23**: 684–695. (*Translated by* E. R. Hope, Directorate of Scientific Information Services, Defense Research Board Canada. Issued Jan. 1964.)

Sander, E.
  1967 Skimmers in the marine aquarium. *Petfish Monthly*, May: 48–51.

Saunders, R. L.
  1962 The irrigation of the gills in fishes. II. Efficiency of oxygen uptake in relation to respiratory flow activity and concentrations of oxygen and carbon dioxide. *Can. J. Zool.*, **40**: 817–862.

Schlieper, C.
  1950 Temperaturbezogene Regulationen des Grundumsatzes bei Wechselwarmen Tieren. *Biol. Zentralbl.*, **69**: 216–227.

Schmalz, R. F., and K. E. Chave
  1963 Calcium carbonate: factors affecting saturation in ocean waters off Bermuda. *Science*, **139**: 1206–1207.

Segedi, R., and W. E. Kelley
  1964 A new formula for artificial sea water. *In* Sea-water systems for experimental aquariums: a collection of papers, J. R. Clark and R. L. Clark (eds.). U.S. Dept. of the Interior, Bureau of Sport Fisheries and Wildlife. Res. Rept. 63. 192 pp.

Shelbourne, J. E.
  1964 The artificial propagation of marine fish. *In* Advances in marine biology, Vol. 2, F. S. Russell (ed.). Academic Press, New York. 274 pp.

Shirahata, S.
  1964 Problems of water quality in food trout production. *Bull. Fac. Fish.* (*Nagasaki Univ.*), **17**: 68–82.

Sieburth, J. M.
  1965 Organic aggregation in sea water by alkaline precipitation of inorganic nucli during the formation of ammonia by bacteria. *Proc. Soc. Gen. Microbiol.*, **41**: xx.

Simkiss, K.
  1964a Variations in the crystalline form of calcium carbonate precipitated from artificial sea water. *Nature*, **201**: 492–493.

Simkiss, K.
  1964b The inhibitory effects of some metabolites on the precipitation of calcium carbonate from artificial and natural sea water. *J. Conseil Intern. Exploration Mer*, **29**: 6–18.

Smith, H. W.
  1929 The excretion of ammonia and urea by the gills of fish. *J. Biol. Chem.*, **81**: 727–742.

Smith, L. O., Jr., and S. J. Cristol
  1966 Organic chemistry. Reinhold, New York. 966 pp.

Stanier, R. Y., M. Doudoroff, and E. A. Adelberg
  1963 The microbial world. 2nd ed. Prentice-Hall, Englewood Cliffs, New Jersey. 753 pp.

Strickland, J. D. H., and T. R. Parsons
    1968   A practical handbook of seawater analysis. *Bull. Fish. Res. Board Can.*, **167**: 311 pp.

Sutcliffe, W. H., E. R. Baylor, and D. W. Menzel
    1963   Sea surface chemistry and Langmuir circulation. *Deep-Sea Res.*, **10**: 233–243.

Thomas, A.
    1915   Effects of certain metallic salts upon fishes. *Trans. Amer. Fish. Soc.*, **44**: 120–124.

Tomlinson, T. G., A. G. Boon, and G. N. A. Trotman
    1966   Inhibition of nitrification in the activated sludge process of sewage disposal. *J. Appl. Bacteriol.*, **29**: 266–291.

Tyler, A. V.
    1966   Some lethal temperature relations of two minnows of the genus *Chrosomus*. *Can. J. Zool.*, **44**: 349–361.

Vaccaro, R. F.
    1965   Inorganic nitrogen in sea water. *In* Chemical Oceanography, Vol. 1, J. P. Riley and G. Skirrow (eds.). Academic Press, New York. 712 pp.

Warren, K. S.
    1962   Ammonia toxicity and pH. *Nature*, **195**: 47–49.

Warren, K. S., and S. Schenker
    1962   Differential effect of fixed acid and carbon dioxide on ammonia toxicity. *Amer. J. Physiol.*, **203**: 903–906.

Weinberger, L. W., D. G. Stephan, and F. M. Middleton
    1966   Solving our water problems—water renovation and re-use. *Ann. N. Y. Acad. Sci.*, **136**: 133–154.

Wells, A. N.
    1935   The influence of temperature upon the respiratory metabolism of the Pacific killifish, *Fundulus parvipinnis*. *Physiol. Zool.*, **8**: 196–227.

Westfall, B. A.
    1945   Coagulation film anoxia in fishes. *Ecology*, **26**: 283–287.

White, G. F., and A. J. Thomas
    1912   Studies on the absorption of metallic salts by fish in their natural habitat. I. Absorption of copper by *Fundulus heteroclitus*. *J. Biol. Chem.*, **11**: 381–386.

Wiens, A. W., and K. B. Armitage
    1961   The oxygen consumption of the crayfish *Orconectes immunis* and *Orconectes nais* in response to temperature and oxygen saturation. *Physiol. Zool.*, **34**: 39–54.

Wolf, K.
    1957   Blue-sac disease investigations: microbiology and laboratory induction. *Progressive Fish-Culturist*, **19**: 14–18.

Wood, J. D.
    1958   Nitrogen excretion in some marine teleosts. *Can. J. Biochem. Physiol.*, **36**: 1237–1242.

Wuhrmann, K.
    1952   La protection des rivieres contre la pollution. *Bull. Centre Belge Etude Doc. Eaux (Liege)*, **15**: 77–85.

Yu, M. L.
    1968   A study on the growth inhibiting factors of zebrafish, *Brachydanio rerio* and blue gourami, *Trichogaster trichopterus*. Ph.D. thesis. Dept. of Biology, New York University.

ZoBell, C. E.
1934   Microbiological activities at low temperatures with particular reference to marine bacteria. *Quart. Rev. Biol.*, **9**: 460–466.
ZoBell, C. E., and H. D. Michener
1938   A paradox in the adaptation of marine bacteria to hypertonic solutions. *Science*, **87**: 328–329.

Index

Absorbance, *see* Extinction (*E*)
Acclimation temperature, *see* Ambient temperature
Acetate buffer, 116
Acetone, 116, 117
Acidic cation exchange resins, 48, 49
Acid water, *see* Soft water
Activated carbon, 41-48
  adsorptive mechanism, 21
  airstripping and, 56
  replacement cost, 47, 59
  as tertiary filtration, 58, 59
Activated sludge, 92
Active uptake, 85, 89
Adenosine triphosphate, 69
Aequidens portalegrensis, see Cichlid
Aeration, conditioned systems, 17
  dissolved organics, 41
  in mixing vats, 97
  of stored water, 112
  synthetic sea water, 95
  of tap water, 6
Aerobic bacteria, 8, 17
Aerobic conditions, 4
Aged tap water, 6, 8
Agglutination, 20
Air, in airstripping, 52-57
  oxygen in, 75
  ozone generation, 56
Air bubbles, in airlifts, 9
  in airstripping, 56
  detritus formation, 20, 21
  life-span, 41
Air compressors, 10, 54, 55
Air diffusers, in airstripping, 54-56, 97
  in mixing vats, 97
  see also Airstones; Baffles; Raschig rings

Airlift pumps, advantages, 9
  with airstrippers, 53, 55
  capacity and efficiency, 10
  with carbon contactors, 44, 45
  for cleaning filter beds, 25
  detritus formation, 20
  in mixing vats, 97, 100
  operating principles, 9-11
  subgravel filters, 9, 12, 13
  surface agitation by, 78
Air line, 12, 97
Airstones, 11, 56, 101
Airstripping, 52-56
  concentration of bacteria by, 20
  in disease treatment tanks, 114
  limitations, 58, 59
  ozonation, 21, 56, 57
  removal of ammonia by, 106
Air-water interface, adsorption of organics, 52
  detritus formation, 20
  diffusion of oxygen, 75
  entry of CO_2, 65
  expulsion of CO_2, 78
  gas exchange, 41
  ozone production, 58
Albuminoid ammonia, 57
Aldehydes, 57
Algal blooms, 50
Algae, 35, 36
Alkali-iodide-azide reagent, 118, 119
Alkaline phosphatase, 69
Alkaline water, *see* Hard water
Alkalinity, 60, 67, 68, 70
Alkalinization, 66; *see also* Microzonal alkalinization
Alkali reserve, *see* Alkalinity
Ambient temperature, 80, 81

133

Amino acids, 3, 85
Ammonia, acceptable levels, 18, 51, 103
    accumulation, 5
    animal loading, 18, 19, 51
    bacterial oxidation, 3, 4, 8, 18, 19
    conditioned systems, 17-19
    conversion and pH, 14
    in disease treatment tanks, 114
    elevation during shipping, 81
    excretion of, 102, 103
    from fermentation, 9
    increase after treating diseases, 15
    liberation during detritus formation, 21
    measurement, 115, 116
    origin in culture water, 102, 103
    oxygen and toxicity, 103, 104, 106
    pH and toxicity, 103
    removal, by airstripping, 53
        by ion exchange resins, 48, 51
        by ozonation, 56
    resistance of the host, 110
    specific effects on animals, 105-107
    synergistic compounds of, 93
    time lags in oxidation, 5, 18, 19
    tolerance by animals, 19
    toxicity, 4, 5, 19, 51, 103-106, 110
    toxic mechanisms, 103, 104
    see also Albuminoid ammonia; Ionized
        ammonia; Un-ionized ammonia
Ammonification, under anaerobic con-
        ditions, 9
    bacterial reduction processes, 65
    in conditioned systems, 18
    definition, 3
    detritus formation during, 21
Ammonium chloride, 103, 106
Ammonium hydroxide, 105, 106
Ammonium soaps, 70
Ammonium sulfate, 115
Anaerobic bacteria, 4, 8, 13
Anaerobic conditions, 4, 9
Animal bones, 44
Animal excretia, 36, 41, 102
Animal loading, carrying capacity, 15-18, 51
    in conditioned systems, 18
    over-compensation, 19
    surface area of filters, 7
Animal metabolism, age of animals, 83
    elements, 85
    ions, 85

oxygen consumption, 78-83
    photoperiod, 83
    production of CO_2, 65
    salinity, 78
    size of animals, 82
    temperature, 78-81
    see also Bacterial metabolism
Animal respiration, 66, 75-83
    active uptake of elements, 89
    and CO_2, 70
    copper poisoning, 92, 93
    subgravel filters, 12
Anions, 49, 89
Anoxic conditions, 106
Anthracite, see Coal
Antibodies, 110
Antibiotics, 14, 15, 114
Aquarium Systems, Inc., 94, 95
Aragonite, 66, 67
Artemia, see Brine shrimp
Artificial sea water, see Synthetic sea water
Atmospheric pressure, 84, 85
Autotrophic bacteria, 3, 4, 18

Backwashing, DE filters, 34-39
    principles of, 28
    rapid sand filters, 23, 26, 28
Bacteria, effects of uv irradiation, 58
    equilibrium with energy sources, 108
    resistance to copper, 92
    see also Aerobic bacteria; Anaerobic
        bacteria; Autotrophic bacteria,
        Heterotrophic bacteria; Nitrifying
        bacteria; Protein decomposing
        bacteria; Starch decomposing
        bacteria
Bacterial metabolism, energy sources, 18
    production of CO_2, 65
    salinity fluctuations, 5
    temperature, 14
    toxic compounds, 15
    see also Animal metabolism
Bacterial respiration, 66, 70
Baffles, in airlift pumping, 11
    in airstrippers, 56
    in sand-pressure filters, 27
Basic anion exchange resins, 48-50
Bausch & Lomb Spectronic 20, 115
Bentonite, 49, 50
Bicarbonate alkalinity, see Alkalinity

Bicarbonate ion, 60, 61, 65-67
Biological filters, 3-21
  activated carbon, 44, 45
  airstripping, 56
  ammonia removal, 106
  attachment of bacteria, 3, 7, 211, 114
  calcareous gravel, 67-72
  carrying capacity, 15-18
  chemical filtration, 40, 58
  circulation, 7, 23, 66
  cleaning, 25, 27, 39
  disease treatment tanks, 114
  sand-pressure filters, 38
  sand-vacuum filters, 38
  slow sand filters, 24
  surface agitation, 75, 78
  surface area, 7, 15, 16, 24, 48
  toxic compounds, 15
Biological oxygen demand (BOD, chemical
      oxygen demand (COD), 42
  definition, 8
  mechanical filtration, 22
  oxygen depletion, 78
  Schmutzdecke, 78
Biomphalaria sudanica, see Snails
Bis reagent, 116
Blood, ammonia toxicity, 103, 106,
  copper, 92
  proteins, 89
  root effect, 81
Blood corpuscles, 106
Bluegill sunfish, 102
Blue gourami, 107
Blue-sac disease, 106
Bluntnose minnow, 92
Body feed, 34
Brackish water, dissolved oxygen, 117, 119
  salinity, 6
  washing gravel, 8
Brass flow meter, 98
Breathing rate, 80, 82, 92
Brine, 97-100
Brine shrimp, 21, 113
Bromine, 84
Buffer, addition to culture water, 72
  calcareous gravel, 67-69, 82
  definition, 60
  magnesium, 68
  outbreaks of disease, 110
  Root effect, 81, 82

  synthetic sea water, 95
  in water, 61
Buffer reagent, 116, 117
Bullhead, 81, 82; *see also* Catfishes

Calcareous gravel, 67, 68, 70-72, 82
Calcite, 66-69
Calcium, 89
Calcium carbonate, 60, 65-69
Calcium carbonate scale, 35, 26
Calcium chloride, 67
Calcium soaps, 70
Calgon, 38
Carbon, 3
Carbonate, 84, 92
Carbonate ion, 60, 61, 65-68
Carbon bed, *see* Carbon contactors
Carbon contactors, 42-44, 46-48, 58, 59
Carbon dioxide (free), acclimation of
      fishes, 82
  bacterial oxidation, 69-71
  buffers, 61
  CO_2 system, 60
  from deamination, 3
  disease resistance, 110
  effects on ammonia, 103
  elevation during shipping, 81
  from fermentation, 9
  use by heterotrophs, 3
  and metabolism, 65
  partial pressure in water, 68, 71, 78
  reduction with lime, 71, 72
  from respiration, 66
  Root effect, 81, 82
  solubility, 65, 66, 70
    of copper, 93
  in water, 75
  water and mineral carbonates, 65, 66, 68
Carbon dioxide system, 60-72
Carbonic acid, 65
Carbon tetrachloride, 116
Carp, conditioning new systems, 19
  immunity in mucus, 111
  respiration and age, 83
  Root effect, 81, 82
  stunting from ammonia, 106
  thermal acclimation, 79
Carrying capacity, 15-17
  activated carbon, 48
  ammonia, 51, 106

buffering, 68
  heterotrophic oxidation, 40
  infectious diseases, 110
  ion exchange resins, 51
  Schmutzdecke, 25
Catfishes, 19; *see also* Bullhead
Cations, active uptake by animals, 89
  affecting hardness, 60
  interference in ion exchange, 49
  mineral carbonate solubility, 67
Central air line, 12
Central cores, backwashing, 34, 35
  description, 28, 29
  precoat, 32
Charcoal, *see* Activated carbon
Chelation, 92
Chemical filtration, 3, 40-59
  clogged filter sleeves, 36
  nutrient removal, 21
  pheromone removal, 107
Chemical oxygen demand (COD), 42, 52
Chloramine-T, 116
Chloride ions, 49
Chlorine, *see* Free chlorine; Residual
    chlorine
Chlorine demand, 37
Chlorinity, 84, 85, 117
Chloroform, 118
Cichlid, 79
Citric acid, 92
Coal, 26, 43
Coconut shell, 44
Coho salmon, 81; *see also* Salmon
Cold acclimation, *see* Cold water
Cold-blooded animals, effects of tempera-
    ture, 78-81
  synthesis of antibodies, 110
Cold water, acclimation of warm-water
    fishes, 79, 80
  activated carbon, 43
  animal loading changes, 18
  definition, 17
  nitrification time lags, 19
Collection chamber in airstripping, 53-55
Colloidal coating, of activated carbon, 44
  of filter sleeves, 29, 34-37
  of ion exchange resins, 49-51
  see also Colloidal organics
Colloidal organics, DE filters, 28, 29, 34,
    39, 40

mechanical filtration, 22
  oxidation by sodium hypochlorite, 37
  rapid sand filters, 40
  removal from filter sleeves, 35
Column filter elements, cleaning, 34, 36-38
  description, 29
Conditioned systems, 17-19
  time factor, 108
  disease in, 110, 114
Contact time, activated carbon, 43, 44, 46
  airstripping, 52, 53, 55
  ozonation, 57
  uv irradiation, 58
Contact tube, in airstripping, 54-56
Copper, biochemical functions, 89
  brass flow meters, 98
  chelation and toxicity, 92
  detritus and toxicity, 92
  effects, of CO_2, 93
    of dissolved oxygen, 92, 93
  as hardness factor, 60
  organics and toxicity, 92
  pH and solubility, 93
  prophylactic levels, 92
  synergistic compounds, 93
  respiratory pigment, 89
  toxicity, to nitrifying bacteria, 17
    of precipitate, 93
  treating diseases, 89, 91, 92
Copper sulfate, 92, 116
Copper sulfate reagent, 116, 117
Corner filters, 44
Counter-current, *see* Airstripping
Crushed coral rock, 68
Crushed oyster shell, 68
Crustaceans, 102
Cryptocaryon, 89
Cuprammonium ions, 93
Cycle runs, clogged filter sleeves, 35
  definition, 28
  DE filters, 34, 35, 39

Daphnia, 113
Deamination, anaerobic conditions, 9
  bacterial reduction processes, 66
  definition, 3
  detritus formation, 21
  new systems, 18
Denitrification, activated carbon, 43
  biological filtration, 3

conditioned systems, 18
definition, 4
nitrate reduction and pH, 71
nitrogen cycle, 5
Density, *see* Specific gravity
Depressant organics, *see* Pheromones
Detergents, 35, 38
Detritus, accumulation, in biological filters, 20-27, 66, 70
    in rapid sand filters, 26, 28
    definition, 20
    formation, 20, 21
    grading and shape of gravel, 23
    improved mechanical efficiency with, 23-26
    influence on heavy metal toxicity, 92
    microbial oxidation of, 24
    removal, from biological filters, 22, 25, 27
        from rapid sand filters, 26, 28
    role in bacterial reduction, 66
    shrinking of filter beds, 25
Diamondback terrapins, 19
Diatomaceous earth (DE), cost, 39
    description and efficiency, 28, 29
    prefiltering raw water, 112
Diatomaceous earth filters, 28-39
    advantages and disadvantages, 39
    before ion exchange beds, 50
    for cleaning biological filters, 39
    clogged filter sleeves, 35-39
    DE pressure, 29, 32, 35, 38
    DE vacuum, 29, 32, 35, 36, 38
    operating costs, 39
    maintenance, 36, 38, 39
    as mechanical filters, 40
    prefiltering natural water, 112, 113
    as secondary filtration, 58
    superchlorination, 113
    surface area, 34
Diatoms, 28
Diffusion, definition, 85
    of organics into activated carbon, 42
    of oxygen into water, 75
Dihydrochloride solution, 116, 117
Direct-current, *see* Airstripping
Disease resistance, *see* Immunity
Diseases, effects of uv irratiation, 58
    transfer of, *see* Parasites
    treatment, 15, 113, 114
    see also Environmental stress; Epizootic

diseases; Heavy metals; Immunity
Dissolved chemical oxygen demand, ion exchange resins, 51
    mineralization, 42, 43
    ozonation, 56
    see also chemical oxygen demand (COD); Dissolved organics
Dissolved organics, activated carbon, 41-43, 48, 58
    agglutination and further adsorption, 20, 21
    airstripping, 52-56
    chemical filtration, 40, 107
    coating of carbonate particles, 66, 68-70
    disease resistance, 110
    foaming, 41
    ion exchange resins, 51, 58
    ozone, 56
Dissolved oxygen, active uptake of elements, 89
    ammonia toxicity, 103, 104, 106
    carrying capacity, 15
    copper toxicity, 93
    disease resistance, 110
    in filter beds, 8, 9, 25
    lead nitrate toxicity, 92
    measurement, 117-119
    partial pressure, 78, 79
    Root effect, 81, 82
    solubility, 79, 80
    toxic metabolites, 102
    in water, 75
Distilled water, in laboratory tests, 115-118
    mixing trace elements, 101
    specific gravity, 84
    toxicity of heavy metals, 93
Divalent ions, 89
Dolomite, 68
Dow Corning silicone sealant, 13

Ectoparasites, 110
Effluent water, 10, 25, 43, 48
Electrical conductivity, 117
Electrostatic attraction, dissolved organics, 20
    particulate matter, 22, 56
    shape of gravel, 23
Elements, 85; *see also* Major salts; Minor salts; Trace salts
Environmental stress, biological filtration, 5

disease, 109-114
pollution, 82
soft water, 71
Enzymes, 85, 89
Epidermis, 106
Epizootic diseases, condition of host, 109
  introduction with natural water, 22
  primary causes, 111
  sterilization, 59
Epoxy paint, 13, 44, 101
Erythrocytes, destruction by ammonia, 106;
    see also Blood; Blood corpuscles
Evaporation, 6, 95
Extinction (E), 115

Fatty acids, interference with gas exchange,
    41
  reaction with ammonia, 70
Fatty alcohols, 41
Fecundity, 102, 107
Fermentation, 10
Fiberglass components, 13
Filter bays, cleaning filter elements, 34, 35
  description, 29
  parasites, 112
Filter cake, 29, 32, 34, 39
Filter elements, clogging, 35-39
  description, 29
  inspection, 38
  parasites, 112
  precoating, 29, 32
  surface area, 34
  see also Column filter elements; Leaf filter
    elements
Filtering velocity, see Turnover rate
Filter plate, see Subgravel filter
Filter sleeves, 29, 35-38
Fishes, ammonia poisoning, 19
  carrying capacity, 15-17
  chelated, 92
  cold-acclimated, 79
  to condition new systems, 19
  culture in synthetic sea water, 95
  diseases, 110
  effects of ammonia, 105, 106
  excretion of ammonia, 102
  heavy metals, 89, 91-93
  immunity, 110, 111
  lead salts, 92, 93
  live food, 113

photoperiod, 83
respiration, 78-83
Root effect, 81, 82
thermal acclimation, 79-81
warm-acclimated, 79
Fish flesh, 17, 36
Fish mucus, affected by heavy metals, 89,
    92, 93, 114
  antibodies, 110
  immunity, 110, 111
Fish oils, 36
Fixed acid, 103
Fixed base, 103
Flexible hose, 44, 45
Float valve, 98
Flow meter, 98
Flow rate, see Turnover rate
Foam fractionation, see Airstripping
Foaming, in sea water, 41
Food input, carrying capacity, 15-17
  to conditioned systems, 18
Formalin, 14, 114
Free chlorine, clogged filter sleeves, 37
  salinity, 84
  superchlorination, of culture systems, 114
    of prefilters, 113
  in tap water, 6
Freezing point depression, 117
Frozen animal flesh, 113
Further adsorption, in detritus formation,
    20

Gases, 75
Gas exchange, 41
General Electric silicone sealant, 13
Gill epithelium, 106
Gill filaments, 105
Gill hyperplasia, 105, 106
Gills, damage by heavy metals, 92
  elimination of ammonia, 102
  oxygen uptake, 79
  parasites, 110
Glacial acetic acid, 116
Glass wool, 44
Goldfish, lead nitrate, 92
  oxygen consumption, 79
  photoperiod, 83
  Root effect, 81
Gravel, bacterial attachment, 3, 7
  bacterial detachment, 7, 8, 25, 28

conditioned systems, 19, 108
depth, 15, 16, 24
in filter beds, 23
grain size, 7, 8, 13, 15, 16, 22-24, 26, 68
ideal grain size, 8, 24
in rapid sand filters, 23
shape, 8, 23, 24
surface area of, 22
surfaces and biological filtration, 6-8
washing, 7, 8
see also Calcareous gravel; Silica gravel;
    Sand
Gravel bed, *see* Biological filters
Groupers, 19
Growth inhibition, 102, 105-108
Growth promotion, 108
Glycerophosphate, 69

Halides, 84
Halogens, 84
Hardness, 60
Hard water, 65, 93
Head loss, in biological filters, 24
in DE pressure filters, 34
in rapid sand filters, 28
Heavy metals, factors affecting toxicity, 92,
    93
fish diseases, 89-91, 114
lethal levels, 89, 91, 92
long-term effects, 91
mucus production, 89, 114
toxicity, 89-93
toxic mechanisms, 92, 93
see also Copper; Lead; Mercury; Zinc
Heterotrophic bacteria, in detritus formation,
    20
fat hydrolysis by, 70
mineralization by, 3, 40, 66, 102
as pathogens, 110
population stabilization, 18
respiration, 70, 71
utilization of ammonia, 18
Host-parasite relationship, 89, 90, 109-114
Hydrazine sulfate, 117
Hydrochloric acid, 49, 50, 68, 117
Hydrogen ions, 4, 65, 67; *see also* pH
Hydrogen sulfide, 9
Hydrometers, 85, 117

Immunity, in mucus, 110, 111

to parasites, 114
toxic metabolites, 102, 105
Infectious agents, *see* Parasites
Influent water, 27, 39, 45, 49, 58
Inorganics, as autotrophic substrates, 3
from deamination, 3
in detritus formation, 20, 21
solutes in sea water, 35
Insecticides, 15
Instant Ocean Synthetic Sea Salts, 94-96,
    99; *see also* Synthetic sea water
International Salt Company, Inc., 97
Invertebrates, ammonia excretion, 102
ammonia poisoning, 19
culture in synthetic sea water, 95
diseases, 110
respiration, 78
respiratory pigments, 89
uptake of dissolved organics, 40
Iodine, 84
Ion exchange resins, 48-51
and airstripping, 56
removal of organic phosphates, 69
Ionization, 4
Ionized ammonia, allowable levels (total
    NH_4^+), 18, 51, 103
from bacterial reduction, 66
measurement (total NH_4^+), 115, 116
Iron, 60
Iron oxide, 35-37

Jackson Turbidity Units (JTU), 42

Ketones, 57

Largemouth bass, 80, 81
Latent infection, 109-111
Lead, 89, 93
Lead nitrate, 92
Leaf filter elements, cleaning, 34, 36-38
description, 29
Limestone, 68; *see also* Calcite
Lipid residues, 41
Live food, 113
Liver, 106

Magnesium carbonate, 65
Magnesium carbonate scale, 35, 36
Magnesium chloride, 67
Magnesium ions, 60, 66-70

Major salts, in synthetic sea water, 93, 95-
    101
Makeup water, brackish systems, 6
    from natural sources, 27
Manganese clogged filter sleeves, 35, 36
    removal with ozonation, 57
Manganous sulfate solution, 118, 119
Manifolds, 29, 34
Map turtles, 19
Marine teleosts, 102
Marine turtles, 19
Mechanical filtration, 3, 22-39
    activated carbon and, 42
    and airstripping, 56
    functions, 22, 40
Mechanical pumps, 26, 29, 32, 46, 97, 98,
    112, 113
Medication, 114
Mercuric chloride, 92
Mercury, 89
Metabolic rate, *see* Animal metabolism;
    Bacterial metabolism
Methane, 9
Methanol, 49, 50
Microorganisms, *see* Suspended micro-
    organisms
Microzonal alkalinization, 21; *see also*
    Alkalinization
Mineral carbonates, Ca-Mg exchange sites,
    67, 70
    in CO_2 system, 60
    precipitation on detritus, 69
    reaction with CO_2 and water, 65
    solubility, 68, 69
    see also Calcium carbonate; Magnesium
    carbonate
Mineralization, ammonia production from,
    102
    by bacteria on activated carbon, 43, 48
    formation of ammonia, 3
    by heterotrophs, 3, 40
    nitrogen cycle, 5
Minnows, effects of ammonia, 80
    as live food, 113
    thermal acclimation, 80
    see also Bluntnose minnow
Minor salts, in synthetic sea water, 95, 96,
    122
Mixing vat *see* Synthetic sea water
Mollusks, 102

Mono reagent, 116
Moray eels, 19
Mucus, *see* Fish mucus
Muriatic acid, 36, 37

Natural sea salts, 93
Natural water, clogged filter sleeves, 35
    as makeup, 27
    prefiltration, 22, 26, 27, 39, 58, 112, 113
Nitrate, acceptable levels, 50, 71
    in conditioned systems, 17
    denitrification of, 4
        on activated carbon, 43
    measurement, 115-117
    and nitrification, 3, 4, 8, 69
    as a "nutrient, " 50
    and pH, 71
    quantities produced, 5
    removal by ion exchange, 48-50
Nitrate formers (nitrite oxidizers), *see*
        Nitrobacter
Nitrification, as an acid-forming reaction, 70
    in biological filtration, 3
    in conditioned systems, 17, 18
    definition, 3
    detritus, 7, 8, 20, 25
    in fresh and sea water, 5
    gravel washing, 7, 8
    interruption of, 114
    nitrogen cycle and, 5
    surface area of filter, 7
    time lags in, 5, 18, 19
    temperature, 14, 19
    in upper gravel layers, 7
Nitrifiers, *see* Nitrifying bacteria; *Nitro-
        bacter; Nitrosomonas*
Nitrifying bacteria, attached, to detritus, 7,
        8, 20, 21
    to gravel surfaces, 8
    chelated copper, 92
    in filter beds, 3-5, 7
    population density, 17-19
    population stabilization, 18, 51
    suspended, 7
Nitrite, animal loading changes and, 18, 19
    conversion and pH, 14
    denitrification, 4, 5
    in fresh and sea water, 5
    levels in conditioned systems, 17-19
    measurement, 115-117

nitrification, 3, 8
time lags in oxidation of, 18, 19
Nitrite formers (Ammonia oxidizers), *see*
    Nitrosomonas
Nitrobacter, 3, 18
Nitrogen, in animal excretia, 102, 103
in denitrification, 4
ion exchange, 49
removal by chemical filtration, 40
Nitrogen cycle, 4
Nitrosomonas, 3, 18
Nitrous oxide, 4
Nucleic acids, 3
Nutrition, 111
Nylon, 29

Odor, in water, 42
Organometallic compounds, 92
Open systems, 35, 46
Organic aggregates, 20, 21
Organic fouling, *see* Colloidal coating
Organics, acids of, 9
  agglutination, 20
  bases of, 3, 9
  chloroform-soluble, 41
  denitrification, 4
  detritus formation, 20, 21
  heavy metal toxicity, 92
  heterotrophic blooms, 71, 72, 110
  nitrogenous, 3
  nonsurface-active, 52
  oxidized by ozone, 57
  polar, 66, 68
  salinity, 84
  saturated, 57, 59
  surface-active, 52, 53, 59
  toxic, 107; *see also* Pheromones
  unsaturated, 57, 59
  see also Colloidal organics; Dissolved
    organics; Phermones
Orthophosphate, 48, 69
Orthotolodine, 36
OTO test, 37, 113
Outside filters, 44
Over-compensation, 19
Oxygen, *see* Dissolved oxygen
Oxygenation, 7-12
Oxygen consumed during filtration (OCF),
    carrying capacity, 15, 16
  definition, 8

nitrification, 8
subgravel filters, 12
Oxygen tension, 8, 81
Ozonation, *see* Ozone
Ozonators, 56, 57
Ozone, 56, 57
  disease treatment tanks, 114
  prefilters, 112
  produced during uv irradiation, 58
  reduction of diseases, 59
  removal of ammonia, 106
  saturated organics, 59
Ozonides, 57

Pacific dogfish shark, 81
Pacific staghorn sculpin, 102
Parasites, effects of heavy metals, 89, 91, 92
  entry with live food, 113
  epizootic outbreaks, 109
  eradication from filter beds, 112
  introduction to culture systems, 111
  transfer, 109, 110, 112
  treatment for, 15
Parasitic arthropods and helminths, control,
    110
  treatment, 114
Parasitic bacteria, control, 110, 111
  effects of uv irradiation, 57, 58
  epizootic outbreaks, 110, 111
  removal with DE, 112
  treatment for, 114
Parasitic protozoans, chronic infections, 111
  control, 89, 110, 111
  effects of uv irradiation, 57, 58
  epizootic outbreaks, 109, 111
  immunity of the host, 110, 111
  removal with DE, 112
  treatment for, 114
Partial water changes, with ozonation, 57
  to reduce nitrate, 71
Particulate matter, activated carbon, 42
  airstripping, 56
  biological filters, 23
  cleaning filter beds, 25
  DE filters, 28, 29, 39, 40, 58
  filter cake porosity, 34
  mechanical filtration, 22
  rapid sand filters, 25, 26, 39, 40
  size removed, by gravel, 26
    with DE, 28

surface area of DE filters, 34
upper gravel layers, 24, 25
Pecan shell, 44
"Per cent" salt, 84
Per cent submergence, in airstripping, 53
    calculation, 10
Per cent transmittance (*T*), 115
pH, acceptable ranges in culturing, 71
    activated carbon, 43, 48
    ammonia removal by airstripping, 53
    ammonia toxicity, 103, 106
    bacterial oxidation, 67, 69, 70
    bacterial reduction, 66
    $CaCO_3$ solubility, 66, 68, 69
    calcareous gravel, 68
    chelated copper, 92
    clogged filter sleeves, 35, 37
    definition, 60, 61
    heavy metal poisoning, 92
    increase in from airstripping, 53
    ion exchange resins, 50
    as a limiting factor, 71, 72
    nitrate accumulation, 71
    nitrification, 14
    ozone stability, 57
    reaction of CO_2 and water, 65
    solubility of copper, 93
    stabilization in synthetic sea water, 95, 97
Phenol, 117
Pheromones, 107, 108
Phosphate, 48-50
Phosphorus compounds, 40, 49
Photoperiod, 83
Physiological stress, *see* Environmental stress
Plastic garbage cans, 94, 97
Plastic screen, 44, 45
Platy, 83
Polypropylene, 28, 35
Polyvalent ions, 89
Polyvinyl chloride, baffles, 11, 56
    pipe and fittings, 44, 45, 60, 61, 97
Potassium, 60, 67, 89
Potassium dichromate, 118
Potassium hydroxide, 118
Potassium iodide, 118
Precoat, amount, 32
    clogged filter sleeves, 36-38
    definition, 29
    superchlorination of prefilters, 113
Precoat pot, 32

Prefilters, *see* Prefiltration
Prefiltration, clogged filter sleeves, 36
    with DE before ion exchange resins, 50
    by filter beds before carbon contactors,
        44, 48
    of natural (raw) water, 22, 39, 112, 113
    by rapid sand before uv sterilizers, 58
Prepared cleansers, 100
Pressure tanks, 27, 29, 32, 38
Primary filtration, 58
Protein decomposing bacteria, 17
Proteins, 85, 89, 92
Protein skimmers, *see* Airstripping
Pyrazolone reagent, 116
Pyridine, 116
Pyrophosphate, 69

Quarantine, 113, 114

Rainbow trout, effects of ammonia, 103,
        104
    excretion of ammonia, 102
    photoperiod, 83
    see also Trout
Rapid sand filters, backwashing, 28
    bacterial populations, 26, 28, 38
    biological filtration, 24, 38
    grades of gravel, 23, 26
    maintenance, 38
    as mechanical filters, 40
    microbial oxidation, 26, 38
    operating costs, 39
    prefiltration of natural water, 112, 113
    sand-pressure, 26, 27, 38, 42, 46
    sand-vacuum, 26, 29, 38
    superchlorination, 112, 113
    surface area, 26
    water-saving techniques with, 27
Raschig rings, 56
Raw water, *see* Natural water
Reagent blank, 115
Reducer reagent, 117
Regeneration, of activated carbon, 44
    of ion exchange resins, 48
Regeneration furnace, 44
Residual chlorine, 112
Respiratory pigments, 89
Root effect, 81, 82
Rubidium, 89
Rusting, 44

Salinity, in ammonia, nitrite, and nitrate
   tests, 115
  animal respiration, 78, 82
  interference in DO test, 119
  measurement, 84, 117
  mineral carbonates, 68
  nitrification, 5, 6, 8
  normal values, 85
  oxygen, 75
Salinometers, 85
Salmon, 105, 106; *see also* Coho salmon
Salts, 84, 89
Sand, nitrifying bacteria, 7, 8, 21
  in rapid sand filters, 26, 36
  see also Gravel
Sand beds, *see* Biological filters
Schmutzdecke, 25
Secondary filtration, 58
Semiclosed systems, 35
Separation chamber, in airstripping, 54, 55
Sewage effluent, *see* Waste water
Side air-inlet, 12
Silica gravel, 68
Silver, 84
Slaked lime, 71
Sliders, 19
Slow sand filters, 24, 26
Sludge treatment, 43
Snails, 108
Snapping turtles, 19
Sodium acetate trihydrate, 116
Sodium azide, 118
Sodium bicarbonate, 71, 100
Sodium bisulfite, 37
Sodium chloride, 5, 48, 97-99
Sodium hydroxide, 49, 50, 116-118
Sodium hypochlorite, 37, 113
Sodium ions, $CaCO_3$ solubility, 67
  as hardness factors, 60
  in ion exchange, 49
Sodium nitrite, 115
Sodium sulfite, 36
Sodium thiosulfate, in DO test, 118, 119
  free chlorine reduction, 37, 113, 114
Soft water, buffering, 68
  and CO_2, 71
  definition, 60
  heavy metals, 93
  pH fluctuations, 70, 71
Sole, 102

Specific gravity, adjustment in synthetic sea
   water, 95-97, 99, 101
  of brine, 100
  effects on respiration, 82
  measurement, 84, 85, 117
  normal values, 6, 108
  range in culturing, 6, 82
Spectrophotometer, *see* Bausch & Lomb
   Spectronic, 20
Stamina, affected by ammonia, 105, 106;
   see also Swimming performance
Standard biological filter, *see* Biological fil-
   ters
Starch, 118, 119
Starch decomposing bacteria, 17
Starry flounder, 102
Steel drum contactor, 44-46
Steinhart Aquarium, 58
Sterilizers, *see* Ozone; Ultraviolet irradia-
   tion (uv)
Sterling Brinomat, 98-100
Steroline Systems Corp., 58
Stickleback, 92
Striped seaperch, 102
Strontium, 89
Subgravel filter, airlift pumping, 9, 12, 13
  carbon contactors, 45
  design and construction, 13
  surface agitation, 78
  surface area of filter bed, 7
Sucker, 82
Sulfanilamide, 117
Sulfates, 49
Sulfuric acid, 36, 118, 119
Superchlorination, 112-114
Surface agitation, copper poisoning, 93
  detritus formation, 20
  expulsion of CO_2, 70, 71, 78, 82
  solubility of oxygen, 75, 78
  turnover rate, 9
Suspended microorganisms, DE filters, 28,
   39
  mechanical filtration, 22
  ozone, 56
  size destroyed by uv, 58
  uv irradiation, 57, 58
Suspended solids, *see* Particulate matter
Swimming performance, 81
Swordtail, 83
Synergistic effects, 92, 93

Synthetic sea water CaCO₃ solubility, 69
  mixing, 94-101
  water-saving techniques, 27
  see also Instant Ocean Synthetic Sea Salts

Tap water, backwashing DE filters, 34
  cleaning mixing vats, 100
  clogged filter sleeves, 36-38
  mixing synthetic sea water, 95, 97, 98
  superchlorination, 114
  washing activated carbon, 44
  washing filter sleeves, 35
Teflon tape, 44
Temperature, activated carbon, 43, 48
  active uptake of elements, 89
  adjustment in synthetic sea water, 96,
    97, 100
  CaCO₃ solubility, 66, 68
  CO₂ solubility, 66
  chlorinity, 84, 85
  cold-water conditions, 17, 19
  disease prevention, 111
  foaming, 41
  heavy metal poisoning, 92, 93
  influence on animal respiration, 78-81,
    83
  nitrification, 14, 19
  ozone stability, 57
  photoperiod, 83
  salinity, 84, 85
  solubility of oxygen, 75, 78-80
  specific gravity, 84, 85
  as a stress factor, 110, 111
  synthesis of antibodies, 110, 111
  toxic metabolites, 102, 104
  warm-water conditions, 17, 19
Tertiary filtration, 58
Thermal acclimation, 79-81; *see also* Cold
    water; Fishes; Warm water
Thermal shock, prevention, 81
  symptoms in fishes, 80
  see also Thermal stress
Thermal stress, 80, 81, 83
Thermometers, 97
Tissue barriers, 103
Tissues, 92
Tobacco smoke, 15
Toluene, 118
Total organic carbon (TOC), 42
Toxic additives, 14, 15

Toxic animal metabolites, 123-131
  epizootic outbreaks of disease, 111
  see also Ammonia; Pheromones
Toxic bacterial metabolites, 8
Trace salts, depletion by ozone, 57
  in synthetic sea water, 93, 95, 96, 101
Trichodinids, 111
Trout, effects of ammonia, 106
  sensitivity to copper, 93
  see also Rainbow trout
Turbidity, activated carbon, 42
  bentonite, 50
  biological filters, 23, 58
  DE filters, 39
  natural water, 112
  rapid sand filters, 22, 25
  uv irradiation, 58
  see also Particulate matter; Suspended
    microorganisms
Turnover rate, airstripping, 53
  ammonia toxicity, 106
  biological filters, 9, 15, 16, 24
  carbon contactors, 42, 46
  carrying capacity, 15, 16
  CO₂ level, 70, 71, 82
  gas exchange, 41, 78
  heavy metal toxicity, 92
  and pH, 66
  rapid sand filters, 25, 26

Ultraviolet irradiation (uv), 57-59, 112
Undergravel filter, *see* Subgravel filter
Un-ionized ammonia, disease resistance, 110
  per cent of total NH₄⁺, 115
  production by heterotrophs, 3, 66
  reaction with CO₂ and water, 66
  toxicity, 53, 103-106
  see also Ammonia; Ionized ammonia
Univalent ions, 89
Urea, 3, 102
Urine, 102

Vaterite, 66, 67
Ventilation, 37
Viruses, 58, 110
Vitamins, 40

Warm acclimation, *see* Warm water
Warm-blooded animals, 102
Warm water, acclimation of cold-water

fishes, 79, 80
activated carbon, 43
animal loading changes, 18
definition, 17
nitrification time lags, 19
turnover rate, 78
Waste water, ammonia removal from, 48, 53
chemical filtration, 41
nitrate removal, 48, 49
orthophosphate removal, 48
and pH, 53
phosphate removal, 48, 49
reduction of bacteria, 56

removal, of COD, 52, 53
of manganese, 57
sand-vacuum filters, 26
treatment with activated carbon, 41-43
Water clarity, 23, 25, 28, 39
Water color, 42
Water softeners, 35
Wood, 43

Zebrafish, 107
Zinc, 89, 93
Zinc ions, 60
Zinc sulfate, 92